DIFFUSE ALGORITHMS FOR NEURAL AND NEURO-FUZZY NETWORKS

DIFFUSE ALGORITHMS FOR NEURAL AND NEURO-FUZZY NETWORKS

With Applications in Control Engineering and Signal Processing

BORIS A. SKOROHOD

ELSEVIER

Butterworth-Heinemann
An imprint of Elsevier
elsevier.com

Butterworth-Heinemann is an imprint of Elsevier
The Boulevard, Langford Lane, Kidlington, Oxford OX5 1GB, United Kingdom
50 Hampshire Street, 5th Floor, Cambridge, MA 02139, United States

British Library Cataloguing-in-Publication Data
A catalogue record for this book is available from the British Library

Library of Congress Cataloging-in-Publication Data
A catalog record for this book is available from the Library of Congress

ISBN: 978-0-12-812609-7

Publisher: Joe Hayton
Acquisition Editor: Sonnini R. Yura
Editorial Project Manager: Ashlie Jackman
Production Project Manager: Kiruthika Govindaraju
Cover Designer: Limbert Matthew

Typeset by MPS Limited, Chennai, India

CONTENTS

LIST OF FIGURES

LIST OF TABLES

PREFACE

The problem of neural and neuro-fuzzy networks training is considered in this book. The author's attention is concentrated on the approaches which are based on the use of a separable structure of plants models—nonlinear with respect to some unknown parameters and linear relating to the others. It may be, for example, a multilayered perceptron with a linear activation function at its output, a radial base neural network, a neuro-fuzzy Sugeno network, or a recurrent neural network, which are widely used in a variety of applications relating to the identification and control of nonlinear plants, time series forecasting, classification, and recognition.

Static neural and neuro-fuzzy networks training can be regarded as a problem of minimizing the quality criterion in respect to unknown parameters included in the description of them for a given training set. It is well-known that it is a complex, multiextreme, often ill-conditioned nonlinear optimization problem. In order to solve its various algorithms, that are superior to the error backpropagation algorithm and its numerous modifications in convergence rate, approximation accuracy and generalization ability have been developed. There are also algorithms that directly take into account separable character of networks structure. Thus, in Ref. [1] the VP (variable projection) algorithm for static separable plants models is proposed. According to this algorithm, the initial optimization problem is transformed into a new problem, but only with relation to nonlinear input parameters. Under certain conditions the stationary points sets of two problems coincide, but at the same time dimensionality decreases, and, as a consequence, there is no need for selecting initial values to linearly incoming parameters. Moreover, the new optimization problem is better conditioned [2—4], and if the same method is used for initial and transformed optimization problems, the VP algorithm always converges after a smaller number of iterations. At the same time, the VP algorithm can be implemented only in a batch mode and, in addition, the procedure of determining partial derivative of modified criteria in respect to the parameters becomes considerably more complicated. The hybrid procedure for a Sugeno fuzzy network training is proposed in Refs. [5,6], that is based on the successive use of the recursive least-square method (RLSM) for determining linearly entering parameters and gradient method for nonlinear ones. The extreme learning machine (ELM) approach is developed in Refs. [7,8]. On the basis of this approach only linearly incoming parameters are trained and nonlinear ones are drawn at random without taking into account a training set. However, it is well-known that this approach can provide quite low accuracy at a relatively small size of the training set. It also should be noted that while the ELM and the hybrid algorithms initialization use the RLS algorithm, it is necessary to select the initial values for the matrix which satisfies the Riccati equation. Moreover, as a priori information about the estimated parameters is absent, its elements are generally put proportional to a large parameter, which may lead to divergence in the case of even linear regression.

The purpose of this book is to present new approaches to training of neural and neuro-fuzzy networks which have a separable structure. It is assumed that in addition to the training set a priori information only about the nonlinearly incoming parameters is given. This information may be obtained from the distribution of a generating sample, a training set, or some linguistic information. For static separable models the problem of minimizing a quadratic criterion that includes only that information is considered. Such a problem statement and the Gauss–Newton method (GNM) with linearization around the latest estimate lead to new online and offline training algorithms that are robust in relation to unknown a priori information about linearly incoming parameters. To be more precise, they are interpreted as random variables with zero expectation and a covariance matrix proportional to an arbitrarily large parameter μ (soft constrained initialization). Asymptotic representations as $\mu \to \infty$ for the GNM, which we call diffuse training algorithms (DTAs), are found. We explore the DTA properties. Particularly the DTAs' convergence in case of the limited and unlimited sample size is studied. The problem specialty is connected with the observation model separable character, and the fact that the nonlinearly inputting parameters belong to some compact set, and linearly inputting parameters should be considered as arbitrary numbers.

It is shown that the proposed DTAs have the following important characteristics:

1. Unlike their prototype, the GNM with a large but finite μ, the DTAs are robust with respect to round-off error accumulation.
2. As in Refs. [1–4] initial values choice for linearly imputing parameters is not required, but at the same time there is no need to evaluate the projection matrix partial derivative.

3. Online and offline regimes can be used.
4. The DTAs are followed with the ELM approach and the hybrid algorithm of the Sugeno neuro-fuzzy network training [6,7], and presented modeling results show that developed algorithms can surpass them in accuracy and convergence rate.

With a successful choice of a priori information for the nonlinear parameters, rapid convergence to one of the acceptable minimum criteria points can be expected. In this regard, the DTAs' behavior analysis at fixed values of the nonlinear parameters, when a separable model is degenerating into a linear regression, is very important. We attribute this to the possible influence of the properties of linear estimation problem on the DTAs. The behavior of the RLSM with soft and diffuse initialization in a finite time interval, including a transition stage, is considered. In particular, the asymptotic expansion for the solution of the Riccati equation, the gain rate in inverse powers of μ, and conditions for the absence of overshoot in the transition phase are obtained. Recursive estimation algorithms (diffuse) as $\mu \to \infty$ not depending on a large parameter μ which leads to the divergence of the RLSM are proposed.

The above-described approach is generalized in the training problem of separable dynamic plant models—a state vector and numerical parameters are simultaneously evaluated using the relations for the extended diffuse Kalman filter (DKF) obtained in this book. It is assumed that in addition to the training set a priori information only on nonlinearly inputting parameters and an initial state vector, which can be obtained from the distribution of a generating sample, is used. Linearly inputting parameters are interpreted as random variables with a zero expectation and a covariance matrix proportional to arbitrarily large parameter μ. Asymptotic relations as $\mu \to \infty$, which describe the extended KF (EKF), are called the diffuse extended KF.

The theoretical results are illustrated with numerical examples of identification, control, signal processing, and pattern recognition problem-solving. It is shown that the DTAs may surpass the ELM and the hybrid algorithms in approximation accuracy and necessary iterations number. In addition, the use of the developed algorithms in a variety of engineering applications, which the author has been interested in at different times, is also described. These are dynamic mobile robot model identification, neural networks-based modeling of mechanical hysteresis deformations, and monitoring of the electric current harmonic components.

The book includes six chapters. The first chapter presents an overview of the known models of objects and results relating to the subject of the book.

The RLSM behavior on a finite interval is considered in Chapter 2, Diffuse Algorithms for Estimating Parameters of Linear Regression. It is assumed that the initial value of the matrix Riccati equation is proportional to a large positive parameter μ. Asymptotic expansions of the Riccati equation solution and the RLSM gain rate in inverse powers of μ are obtained. The limit recursive algorithms (diffuse) as $\mu \to \infty$ not depending on a large parameter μ which leads to the RLSM divergence are proposed and explored. The theoretical results are illustrated by examples of solving problems of identification, control, and signal processing.

In Chapter 3, Statistical Analysis of Fluctuations of Least Squares Algorithm on Finite Time Interval, properties of the bias, the matrix of second-order moments, and the normalized average squared error of the RLSM on a finite time interval are studied. It is assumed that the initial condition of the Riccati equation is proportional to the positive parameter μ and the time interval includes an initialization stage. Based on the Chapter 2, Diffuse Algorithms for Estimating Parameters of Linear Regression results, asymptotic expressions for these quantities in inverse powers of μ for the soft initialization and limit expression for the diffuse initialization are obtained. It is shown that the normalized average squared error of estimation can take arbitrarily large but bounded values as $\mu \to \infty$. The conditions are expressed in terms of signal/noise ratio under which overshoot does not exceed the initial value (conditions for the absence of overshoot).

Chapter 4, Diffuse Neural and Neuro-Fuzzy networks Training Algorithms deals with the problem of multilayer neural and neuro-fuzzy networks training with simultaneous estimation of the hidden and output layer parameters. The hidden layer parameters probable values and their possible deviations are assumed to be known. A priori information about the output layer weights is absent and in one initialization of the GNM they are assumed to be random variables with zero expectations and a covariance matrix proportional to the large parameter and in the other option either unknown constants or random variables with unknown statistical characteristics. Training algorithms based on the GNM with linearization about the latest estimate are proposed and studied. The theoretical results are illustrated with the examples of pattern recognition, and identification of nonlinear static and dynamic plants.

The estimation problem of the state and the parameters of the discrete dynamic plants in the absence of a priori statistical information about initial conditions or its incompletion is considered in Chapter 5, Diffuse Kalman Filter. Diffuse analogues of the Kalman filter and the extended Kalman filter are obtained. As a practical application, the problems of the filter constructing with a sliding window, observers restoring state in a finite time, recurrent neural networks training, and state estimation of nonlinear systems with partly unknown dynamics are considered.

Chapter 6, Applications of Diffuse Algorithms provides examples of the use of diffuse algorithms for solving problems with real data arising in various engineering applications. They are the mobile robot dynamic model identification, hysteresis mechanical deformations modeling on the basis of neural networks, and electric current harmonic components monitoring.

The author expresses deep gratitude to Head of Department Y.B. Ratner from Marine Hydrophysical Institute of RAS, Department of Marine Forecasts and Professor S.A. Dubovik from Sevastopol State University, Department of Informatics and Control in Technical Systems for valuable discussions, and to his wife Irina for help in preparation of the manuscript.

Boris Skorohod,
Sevastopol, Russia
January 2017

CHAPTER 1

Introduction

Contents

1.1 SEPARABLE MODELS OF PLANTS AND TRAINING PROBLEMS ASSOCIATED WITH THEM

1.1.1 Separable Least Squares Method

Let us consider an observation model of the form

$$y_t = \Phi(z_t, \beta)\alpha, \quad t = 1, 2, \ldots, N, \tag{1.1}$$

where $z_t = (z_{1t}, z_{2t}, \ldots, z_{nt})^T \in R^n$ is a vector of inputs, $y_t = (y_{1t}, y_{2t}, \ldots, y_{mt})^T \in R^m$ is a vector of outputs, $\alpha = (\alpha_1, \alpha_2, \ldots, \alpha_r)^T \in R^r$, $\beta = (\beta_1, \beta_2, \ldots, \beta_l)^T \in R^l$ are vectors of unknown parameters, $\Phi(z_t, \beta)$ is an $m \times r$ matrix of given nonlinear functions, R^l is the space of vectors of length l, $(\cdot)^T$ is the matrix transpose operation, and N is a sample size.

The vector output y_t depends linearly on α and nonlinearly on β. This model is called the separable regression (SR) [1]. If the vector β is known, then Eq. (1.1) is transformed into a linear regression.

Let there be given the set of input–output pairs $\{z_t, y_t\}$, $t = 1, 2, \ldots, N$ and the quality criteria

Diffuse Algorithms for Neural and Neuro-Fuzzy Networks.
DOI: http://dx.doi.org/10.1016/B978-0-12-812609-7.00001-9

$$J_N(\alpha,\beta)=1/2\sum_{k=1}^{N}(y_k-\Phi(z_k,\beta)\alpha)^T(y_k-\Phi(z_k,\beta)\alpha)= \\ =1/2||Y-F(\beta)\alpha||^2=1/2e^T(\alpha,\beta)e(\alpha,\beta), \tag{1.2}$$

where $Y=(y_1^T,\ldots,y_N^T)^T$, $F(\beta)=(\Phi^T(z_1,\beta),\ldots,\Phi^T(z_N,\beta))^T$, $e(\alpha,\beta)=Y-F(\beta)\alpha$, and $||\cdot||$ is the Euclidean vector norm.

It is required to find the parameters α and β from the minimum $J_N(\alpha,\beta)$:

$$(\alpha^*,\beta^*)=\arg\min J_N(\alpha,\beta),\quad \alpha\in R^r,\quad \beta\in R^l.$$

To solve this problem we can use any numerical optimization method. For example, the iterative Gauss–Newton method (GNM) oriented on the solution of nonlinear problems with a quadratic criterion of quality or its modification the Levenberg–Marquardt algorithm

$$x_{i+1}=x_i+[J^T(x_i)J(x_i)+\mu I_{r+l}]^{-1}J^T(x_i)e(x_i),\quad i=1,2,\ldots,M,$$

where $x=(\beta^T,\alpha^T)^T\in R^{l+r}$, $J(x_i)$ is the Jacobi matrix of residues $e(x)=Y-F(\beta)\alpha$, I_m is the identity $m\times m$ matrix, $\mu>0$ is a parameter, and M is the number of iterations.

However, methods that take into account the specifics of the problem (the linear dependence of the observed output of some parameters) can be more effective. In Ref. [1] a separable least squares method (LSM) to estimate the parameters included in the description Eq. (1.1) is proposed. The idea of the method is as follows. For a given β the value of the parameter α is defined as the solution of the linear problem by the LSM

$$\alpha=F^+(\beta)Y, \tag{1.3}$$

where $(\cdot)^+$ is the pseudoinverse matrix of the corresponding matrix and α is the solution with minimum norm. Substituting α in Eq. (1.2), we come to a new nonlinear optimization problem, but only with respect to β

$$\min\tilde{J}_N(\beta)=1/2\min||(I_{Nm}-F(\beta)F^+(\beta))Y||^2=1/2\min||P_F(\beta)Y||^2,\ \beta\in R^l, \tag{1.4}$$

where $P_F(\beta)=I_{Nm}-F(\beta)F^+(\beta)$ is the projection matrix. The value of α is determined by substitution of the obtained optimal value β in Eq. (1.3).

The following statement shows that the above two-step algorithm, under certain conditions, does minimize the initial performance criterion Eq. (1.2), allowing you to reduce the number of parameters from $r+l$ to l.

Theorem 1.1([1]).

Assume that the matrix $F(\beta)$ has a constant rank over an open set $\Omega \subset R^l$.

1. If $\beta^* \in \Omega$ is a minimizer of $\tilde{J}_N(\beta)$ and $\alpha^* = F^+(\beta^*)Y$, then (α^*, β^*) is also a minimizer of $J_N(\alpha, \beta)$.
2. If (α^*, β^*) is a minimizer of $J_N(\alpha, \beta)$ for $\beta \in \Omega$, then β^* is a minimizer of $\tilde{J}_N(\beta)$ in Ω and $\tilde{J}_N(\beta^*) = J_N(\alpha^*, \beta^*)$. Furthermore, if there is a unique α among the minimizing pairs of $J_N(\alpha, \beta)$, then α must satisfy $\alpha = F^+(\beta)Y$.

For determination of β from the minimum condition $\tilde{J}_N(\beta)$, and in the minimization of $J_N(\alpha, \beta)$, any numerical methods can be used. However, the criterion $\tilde{J}_N(\beta)$ has a much more complicated structure compared to $J_N(\alpha, \beta)$, so it is important to have an efficient procedure for the Jacobian finding of matrix residues $e(\beta) = P_F(\beta)Y$. In Ref. [9] the following analytical representation for the Jacobi matrix columns was obtained

$$\{J\}_j = -[P_F(\beta)\partial F(\beta)\partial\beta_j F^-(\beta) + (P_F(\beta)\partial F(\beta)\partial\beta_j F^-(\beta))^T]Y, \quad (1.5)$$

where $F^-(\beta)$ is generalized pseudoinverse matrix of $F(\beta)$, which satisfies the conditions

$$F(\beta)F^-(\beta)F(\beta) = F(\beta), \quad (F(\beta)F^-(\beta))^T = F(\beta)F^-(\beta).$$

Although the criterion $\tilde{J}_N(\beta)$ seems to be more complicated than $J_N(\alpha, \beta)$, in Ref. [2] it is noted that the number of arithmetic operations required for the implementation of the separable LSM based on Eq. (1.5) is not more than in the GNM to be used for minimization of $J_N(\alpha, \beta)$.

In addition to the reduction of the dimension of the estimated parameter vector, the separable LSM decreases the conditionality matrix in the GNM in comparison with its prototype [4]. This may reduce the number of iterations required to obtain solutions with a given accuracy. Note also that using the LSM to minimize $\tilde{J}_N(\beta)$ it is not required to have a priori information about the incoming linearly parameters which are interpreted as unknown constants.

1.1.2 Perceptron With One Hidden Layer

Consider a neural network (NN) with a nonlinear activation function (AF) in the hidden layer and a linear in the output layer [10]

$$y_{it} = \sum_{k=1}^{p} w_{ik}\sigma\left(\sum_{j=1}^{q} a_{kj}z_{jt} + b_k\right), \quad i = 1, 2, \ldots, m, \quad t = 1, 2, \ldots, N, \tag{1.6}$$

where z_{jt}, $j = 1, 2, \ldots, q$, $t = 1, 2, \ldots, N$ are inputs, y_{it}, $i = 1, 2, \ldots, m$, $t = 1, 2, \ldots, N$, are outputs, a_{kj}, b_k, $k = 1, 2, \ldots, p$, $j = 1, 2, \ldots, q$ are weights and biases of the hidden layer, w_{ik}, $i = 1, 2, \ldots, m$, $k = 1, 2, \ldots, p$ are weights of the output layer, and $\sigma(x)$ is an AF of the hidden layer.

It is usually assumed that the AF is the sigmoid or the hyperbolic tangent function

$$\sigma(x) = (1 + \exp(-x))^{-1}, \quad \sigma(x) = (\exp(x) - \exp(-x))/(\exp(x) + \exp(-x))$$

which are continuously differentiable and have simple expressions for the derivatives.

In applications associated with use of the NN, one of the basic assumptions is that with them a sufficiently wide class of nonlinear systems can be described. It is well known that if the AF is selected as the sigmoid or the hyperbolic tangent function, the NN Eq. (1.6) can approximate with any accuracy continuous functions on compact sets, provided there is a sufficiently large number of neurons in the hidden layer [11,12].

Using vector–matrix notations, we obtain more compact representation for the NN

$$y_t = W(\sigma(a_1 z_t), \ldots, \sigma(a_p z_t))^T, \quad t = 1, 2, \ldots, N, \tag{1.7}$$

where

$$y_t = (y_{1t}, y_{2t}, \ldots, y_{mt})^T \in R^m, \quad z_t = (z_{1t}, z_{2t}, \ldots, z_{qt}, 1)^T \in R^{q+1},$$

$$W = (w_1, w_2, \ldots, w_m)^T \in R^{m \times p}, \quad w_i = (w_{i1}, w_{i2}, \ldots, w_{ip})^T, \quad i = 1, 2, \ldots, m,$$

$$a_k = (a_{k1}, a_{k2}, \ldots, a_{kq}, b_k), \quad k = 1, 2, \ldots, p.$$

The NN Eq. (1.6) is a special case of the SR for which the weights of the output layer enter linearly and the weights and the biases of the hidden layer enter nonlinearly. Introducing notations

$$\begin{gathered}\alpha = (w_1^T, w_2^T, \ldots, w_m^T)^T \in R^{mp}, \quad \beta = (a_1^T, a_2^T, \ldots, a_p^T)^T \in R^{(q+1)p}, \\ \Phi(z_t, \beta) = I_m \otimes \Sigma(z_t, \beta),\end{gathered} \tag{1.8}$$

where $\Phi(z_t, \beta)$ is an $m \times mp$ matrix, $\Sigma(z_t, \beta) = (\sigma(a_1 z_t), \sigma(a_2 z_t), \ldots, \sigma(a_p z_t))$, and $\otimes$ is the direct product of two matrices, we obtain the representation of the NN Eq. (1.6) in the form Eq. (1.1).

Given a training set $\{z_t, y_t\}$, $t = 1, 2, \ldots, N$, the training of the NN Eq. (1.6) can be considered as the problem of the mean-square error minimizing in respect to unknown parameters (weights and biases) occurring in its description. The backpropagation algorithm (BPA) is successfully used to train Eq. (1.6). At the same time, a slow convergence rate inherent in the BPA is well known that considerably complicates or makes the use of the algorithm in complicated problems practically impossible. In numerous publications devoted to the NNs, different learning algorithms are proposed whose convergence rate and obtained approximation accuracy exceed those of the BPA. In Refs. [13,14], Levenberg–Marquardt and quasi-Newtonian methods are applied that use information on the matrix of second-order derivatives of a quality criterion. In Refs. [15–18], training algorithms are based on the extended Kalman filter (EKF) using an approximate covariance matrix of estimation error. In Refs. [3,4,19–21], for the NNs with the linear AF in the output layer, algorithms are presented that take into account a separable network structure. According to the variable projection algorithm [1], the initial optimization problem is transformed into a problem equivalent with respect to its parameters that enter nonlinearly (weights and biases of the hidden layer). In this case, both the problem dimension and conditionality decrease, which makes it possible to decrease the number of iterations for obtaining a solution. At the same time, this approach can be used only in batch mode and even additively entering measurement errors of the NN outputs enter nonlinearly in the transformed criterion. Moreover, the procedure of partial derivatives determination of the criterion in respect to parameters becomes considerably more complicated. In Refs. [7,22], the extreme learning machine (ELM) algorithm is proposed with the help of which only linearly entering parameters (weights of the output layer) are trained, and nonlinear ones are chosen randomly without taking into account the training set. This decreases learning time but can lead to low approximation accuracy. The NN trained in this way is called functionally connected [23] and it may approximate any continuous functions on compact sets provided that $m = 1$.

1.1.3 Radial Basis Neural Network

For a radial basis neural network (RBNN), relationships connecting its inputs and outputs are specified by the expressions [10]

$$y_{it} = \sum_{k=1}^{p} w_{ik}\varphi(b_k||z_t - a_k||^2) + w_{i0}, \quad i = 1, 2, \ldots, m, \quad t = 1, 2, \ldots, N \tag{1.9}$$

where $z_t = (z_{1t}, z_{2t}, \ldots, z_{nt})^T \in R^n$ is a vector of inputs y_{it}, $i = 1, 2, \ldots, m$, $t = 1, 2, \ldots, N$ are outputs, $a_k \in R^n$, $b_k \in R^+$, $k = 1, 2, \ldots, p$ are centers and scaled factors, respectively, R^+ is set of positive real numbers, $w_{ik}, i = 1, 2, \ldots, m$, $k = 1, 2, \ldots, p$ are weights of the output layer, $w_{i0}, i = 1, 2, \ldots, m$ are biases, and $\varphi(\cdot)$ is a basis function.

If the basic function is the Gauss function

$$\varphi(x) = \exp(-(x-a)^2/b),$$

then the RBNN has a universal approximating property [10].

Given a training set $\{z_t, y_t\}$, $t = 1, 2, \ldots, N$, the training of the RBNN Eq. (1.9) can be considered as the problem of minimizing the mean-square error in respect to unknown parameters (weights, centers, and scaled factors) occurring in its description.

Using vector–matrix notations, we obtain following more compact representation for the RBNN

$$y_t = W(1, \varphi(b_1||z_t - a_1||^2), \ldots, \varphi(b_p||z_t - a_p||^2))^T, \quad t = 1, 2, \ldots, N, \tag{1.10}$$

where $y_t = (y_{1t}, y_{2t}, \ldots, y_{mt})^T \in R^m$ is vector of outputs, $W = (w_1, w_2, \ldots, w_m)^T \in R^{m \times (p+1)}$, $w_i = (w_{i0}, w_{i1}, \ldots, w_{ip})^T$, $i = 1, 2, \ldots, m$.

It is seen that the perceptron with one hidden layer and the RBNN have a similar structure—one hidden layer and one output layer. Moreover, the RBNN is a SR in which linearly entering parameters $w_{ik}, i = 1, 2, \ldots, m$, $k = 1, 2, \ldots, p$ are weights of the output layer and nonlinearly entering ones are centers a_k and scaled factors b_k, $k = 1, 2, \ldots, p$.

Introducing notations

$$\alpha = (w_1^T, w_2^T, \ldots, w_m^T)^T \in R^{m(p+1)}, \ \beta = (b_1, b_2 \ldots, b_p, a_1^T, a_2^T, \ldots, a_p^T)^T \in R^{(n+1)p},$$

$$\Phi(z_t, \beta) = I_m \otimes \Sigma(z_t, \beta),$$

$$\Sigma(z_t, \beta) = (1, \varphi(b_1||z_t - a_1||^2), \ldots, \varphi(b_p||z_t - a_p||^2), \tag{1.11}$$

where $\Phi(z_t, \beta)$ is an $m \times m(p+1)$ matrix, we obtain the representation Eq. (1.9) in the form Eq. (1.1).

There are two approaches to training of the RBNN [10,24]. In the first one, the parameters are determined by using the methods of the nonlinear parametric optimization, such as, for example, the Levenberg–Makvardt algorithm and the EKF requiring significant computational effort. The second approach is based on the separate determination of parameters. At first centers and scaled factors are determined and then weights of the output layer. This leads to a substantial reduction in training time in comparison with the first approach. Indeed, if centers and scaled factors of the RBNN have already been found, the problem of weights determination of the output layer is reduced to a linear problem of the LSM. In Ref. [10] it is proposed to choose the centers values randomly from the training set, and scaled factors are assumed to be equal to the maximum distance between centers. In Ref. [22] parameters of the hidden layer it is proposed to choose randomly from their domains of the definition. It is shown that trained the RBNN is a universal approximator under fairly general assumptions about the basic functions and $m = 1$. However, speeding up the learning process may lead, as in the case of the two-layer perceptron, to divergence.

1.1.4 Neuro-Fuzzy Network

One of the main methods of constructing fuzzy systems is the use of a neuro-fuzzy approach [5]. Its basic idea is that in addition to the knowledge of experts, lying in the knowledge base (rules, linguistic variables, membership functions), numerical information representing a set of input values and the corresponding output values is used. It is an introduced criterion (usually a quadratic one) which determines error between the experimental data and the values of the neuro-fuzzy system (NFS) outputs. Further, the membership function (MF) and outputs are parameterized and a nonlinear optimization problem with respect to the selected parameters is solved in some way. Motivation to use fuzzy rules jointly with experimental data is that, unlike the perceptron, they are easily interpreted and expanded. At the same time, fuzzy models as well as the multilayered perceptron allow to approximate with arbitrary precision any continuous function defined on a compact set. The effectiveness of the resulting fuzzy system largely depends on the optimization algorithm. Let us consider in more detail models of the NFS and known algorithms for their training.

Let a knowledge base in the Sugeno form be specified if (z_{1t} is A_{1j}) and if (z_{2t} is A_{2j}) and ... if (z_{nt} is A_{nj}) then $y_t = a_j^T z_t + b_j$,

$$j = 1, 2, \ldots, m, \tag{1.12}$$

where A_{ij}, $i = 1, 2, \ldots, n$, $j = 1, 2, \ldots, m$ are fuzzy sets with parameterized MFs $\mu_{ij}(z_{it}, c_{ij})$, c_{ij}, a_j, b_j, $i = 1, 2, \ldots, n$, $j = 1, 2, \ldots, m$, are vectors of unknown parameters, y_t, $t = 1, 2, \ldots, N$ is a scalar output, and $z_t = (z_{1t}, z_{2t}, \ldots, z_{nt})^T \in R^n$ is a vector of inputs.

Then the relationship for the output determined according to m rules Eq. (1.12) is represented by the expression [25]

$$y_t = \frac{\sum_{i=1}^{m} \prod_{j=1}^{n} \mu_{ij}(z_{jt}, c_{ij})(a_i^T z_t + b_i)}{\sum_{i=1}^{m} \prod_{j=1}^{n} \mu_{ij}(z_{jt}, c_{ij})}, \quad t = 1, 2, \ldots, N \tag{1.13}$$

which is a SR. In Eq. (1.13) the MF and consequences parameters enter nonlinearly and linearly, respectively.

The representation Eq. (1.1) follows from Eq. (1.13) with

$$\alpha = (a_1^T, a_2^T, \ldots, a_m^T, b_1, b_2, \ldots, b_m)^T \in R^{m(m+1)}, \quad \beta = (c_{11}^T, c_{21}^T, \ldots, c_{nm}^T)^T,$$
$$\Phi(z_t, \beta) = (q_1(\beta) z_t^T, q_2(\beta) z_t^T, \ldots, \; q_m(\beta) z_t^T, q_1(\beta), q_2(\beta), \ldots, q_m(\beta))^T,$$

$$q_i(\beta) = \frac{\prod_{j=1}^{n} \mu_{ij}(z_{jt}, c_{ij})}{\sum_{i=1}^{m} \prod_{j=1}^{n} \mu_{ij}(z_{jt}, c_{ij})}, \quad i = 1, 2, \ldots, m.$$

There are various algorithms for estimation of the MF parameters and the NFS parameters entering in rules consequences. So in Refs. [5,6] the system ANFYS is presented. It includes the BPA and hybrid learning procedure, integrating the BPA and the recursive LSM (RLSM). In Ref. [26] the EKF is used for training with the triangular MFs. Algorithms of the cluster analysis and the evolutionary programming are developed in Refs. [5,27,28]. In many cases, these approaches have serious problems due to the absence of convergence or slow convergence and, besides, only to a local minimum. It is known that convergence depends on the used algorithm and on the selected initial conditions for parameters. At the same time, the structure of the NFS is such that a priori information in respect to consequences parameters is absent and, therefore, it is not clear how reasonably to initialize the chosen algorithm.

1.1.5 Plant Models With Time Delays

One of the standard approaches to build dynamic system models consists in the interpretation of z_t in Eq. (1.1) as a regressor defined on the measured values of inputs and outputs to the current time [10]. Suppose that at time $t = 1, 2, \ldots, N$ input $u_t \in R^n$ and output $y_t \in R^m$ values of some system are measured. Let us introduce the regressors vector

$$Z_t = (y_{t-1}, y_{t-2}, \ldots, y_{t-a}, u_{t-d}, u_{t-d-1}, \ldots, u_{t-d-b}),$$

where a, b, d are some positive integer numbers. A nonlinear model of autoregressive moving average connecting input Z_t and output y_t of the system can be represented in the form

$$y_t = F(Z_t, \theta) + \xi_t, \quad t = 1, 2, \ldots, N \tag{1.14}$$

where $F(\cdot, \cdot)$ is a given nonlinear function, θ is vector of unknown parameters, $\xi_t \in R^m$ is a random process that has uncorrelated values, zero expectation, covariance matrix $V_t = E[\xi_t \xi_t^T]$ and characterizes output measurement errors and approximation errors.

The model Eq. (1.14) predicts one step in the behavior of the system. If more than one step is to be predicted we can use Eq. (1.14) and $\hat{y}_t$ to obtain $\hat{y}_{t+1}$. This procedure can be repeated h times to predict h steps ahead. We can also directly build the model of the form

$$y_{t+h} = F(Z_t^h, \theta) + \xi_t, \quad t = 1, 2, \ldots, N \tag{1.15}$$

to predict h steps ahead, where

$$Z_t^h = (y_{t-1}, y_{t-2}, \ldots, y_{t-a}, u_{t+h-d}, u_{t+h-d-1}, \ldots, u_{t-d-b}).$$

Advantages and disadvantages of each of these approaches are well known [29]. We can use the NN or the NFS to choose $F(\cdot, \cdot)$ in Eq. (1.14) and we are interested in a model of the form Eq. (1.14) that can be reduced to a SR Eq. (1.1).

1.1.6 Systems With Partly Unknown Dynamics

Consider the cases where the mathematical description of the studied system is given in the form of a state-space model that is not complete. This may be, for example, if the right sides are defined up to unknown parameters or the right sides of the prior model are different from the real.

Let a system model have the form

$$x_{t+1} = A_t(\theta)x_t + B_t(\theta)u_t, \tag{1.16}$$

$$y_t = C_t(\theta)x_t + D_t(\theta)u_t, \quad t = 1, 2, \ldots, N \tag{1.17}$$

where $x_t \in R^n$ is a state vector, $y_t \in R^m$ is a vector of measured outputs, $u_t \in R^l$ is a vector of measured inputs, $\theta_t \in R^r$ is a vector of unknown parameters, and $A_t(\theta)$, $B_t(\theta)$, $C_t(\theta)$, $D_t(\theta)$ are given matrix functions of corresponding dimensions. It is required using observations $\{u_t, y_t\}$, $t = 1, 2, \ldots, N$ to estimate θ.

Let us show that this problem can be reduced to the estimation of regression parameters (x_1, θ), where x_1 enters linearly and θ nonlinearly. Indeed, from Eq. (1.16) we find

$$x_t = H_{t,1}(\theta)x_1 + \sum_{i=1}^{t-1} H_{t,i+1}(\theta)\, B_i(\theta)u_i, \tag{1.18}$$

where the transition matrix $H_{t,s}(\theta)$ is a solution of the matrix equation $H_{t+1,s}(\theta) = A_t(\theta)H_{t,s}(\theta)$, $H_{s,s}(\theta) = I_n$, $t \geq s$, $t = 1, 2, \ldots, N$.

Substituting Eq. (1.18) in the observation Eq. (1.17) gives

$$y_t = C_t(\theta)H_{t,1}(\theta)x_1 + C_t(\theta)\sum_{i=1}^{t-1} H_{t,i+1}(\theta)\, B_i(\theta)u_i + D_t(\theta)u_t =$$
$$= \Phi_t(\theta)x_1 + \varphi_t(\theta), \quad t = 1, 2, \ldots, N.$$

Suppose that the quality criterion takes the form

$$J(x_1, \theta) = 1/2\|Y - \Phi(\theta)x_1 - \varphi(\theta)\|^2, \tag{1.19}$$

where

$$Y = (y_1^T, y_2^T, \ldots, y_N^T)^T, \quad \Phi(\theta) = (\Phi_1^T(\theta), \Phi_2^T(\theta), \ldots, \Phi_N^T(\theta))^T,$$

$$\varphi(\theta) = (\varphi_1^T(\theta), \varphi_2^T(\theta), \ldots, \varphi_N^T(\theta))^T.$$

To determine the vector (x_1, θ) from the minimum condition Eq. (1.19), we can use the idea of the separable LSM finding x_1 with the help of the LSM for a fixed θ. As result we get

$$x_1 = \Phi^+(\theta)(Y - \varphi(\theta)).$$

Substitution of this expression in Eq. (1.19) leads to the criterion

$$\tilde{J}(\theta) = 1/2\|Y - \Phi(\theta)\Phi^+(\theta)(Y - \varphi(\theta)) - \varphi(\theta)\|^2$$

and the new optimization problem, but only relatively θ.

In Refs. [30,31] conditions are obtained under which we can apply Theorem 1.1 for stationary systems Eq. (1.16). Note that this work uses a different parameterization, leading to SR of standard form. In Ref. [31], a recursive algorithm to minimize the criterion $\tilde{J}_N(\beta)$ is proposed.

Consider now a stochastic nonlinear discrete system of the form

$$x_{t+1} = f_t(x_t, u_t) + w_t, \tag{1.20}$$

$$y_t = h_t(x_t, u_t) + \xi_t, \quad t = 1, 2, \ldots, N, \tag{1.21}$$

where $x_t \in R^n$ is a state vector, $y_t \in R^m$ is a vector of measured outputs, $u_t \in R^l$ is a vector of measured inputs, $f_t(x_t, u_t)$, $h_t(x_t, u_t)$ are a priori defined vector functions, and w_t, ξ_t are random perturbations with known statistical properties.

Let the functions $f_t(x_t, u_t)$ and $h_t(x_t, u_t)$ differ from actual

$$\tilde{f}_t(x_t, u_t) = f_t(x_t, u_t) + \varepsilon_t,$$

$$\tilde{h}_t(x_t, u_t) = h_t(x_t, u_t) + \eta_t,$$

where ε_t, η_t are some unknown functions. In Refs. [32–34], the functions ε_t, η_t are proposed to approximate with the help of two multilayer perceptrons and to estimate their parameters simultaneously with the state x_t.

1.1.7 Recurrent Neural Network

Consider a recurrent neural network (RNN) of the form [10]

$$x_t = \begin{pmatrix} \sigma(a_1^T x_{t-1} + b_1^T z_t + d_1) \\ \cdots \\ \sigma(a_q^T x_{t-1} + b_q^T z_t + d_q) \end{pmatrix}, \tag{1.22}$$

$$y_t = \begin{pmatrix} c_1^T x_t \\ \cdots \\ c_m^T x_t \end{pmatrix}, \quad t = 1, 2, \ldots, N, \tag{1.23}$$

where $x_t \in R^q$ is a state vector, $z_t \in R^n$ is a vector of measured inputs, $y_t \in R^m$ is a vector of measured outputs, and $a_i \in R^q$, $b_i \in R^n$, $d_i \in R^1$, $c_i \in R^q$, $\sigma(x)$ is the sigmoid or the hyperbolic tangent function.

Or in more compact form

$$x_t = \sigma(Ax_{t-1} + Bz_t + d), \tag{1.24}$$

$$y_t = Cx_t, \quad t = 1, 2, \ldots, N, \tag{1.25}$$

where $A \in R^{q \times q}$, $B \in R^{q \times n}$, $d \in R^q$, $C \in R^{m \times q}$.

It is required with the help of observations $\{z_t, y_t\}$, $t = 1, 2, \ldots, N$ to estimate A, B, C, d.

Note two important properties of the RNN for us. Firstly, it can approximate with arbitrary accuracy solutions of nonlinear dynamical systems on finite intervals and compact subsets of the state [35]. Secondly, the RNN Eqs. (1.22 and 1.23) belongs to the class of separable models. Indeed, A, B, C, d enter in its description nonlinearly and C enters linearly. In Ref. [10], training algorithms of the RNN are given. These are a generalization of the BPA and the EKF.

1.1.8 Neurocontrol

In Fig. 1.1, as an example, a functional scheme of the supervisory control system [36] with the inverse dynamics is shown. We assume that the inverse model has the form of the SR

$$u_t^s = \Phi(Z_t, U_{t-1}, \beta)\alpha, \quad t = 1, 2, \ldots, N, \tag{1.26}$$

where $Z_t = (z_t, z_{t-1}, \ldots, z_{t-a})^T$, $U_{t-1} = (u_{t-1}^s, u_{t-2}^s, \ldots, u_{t-b}^s)^T$. The desired trajectory is used as input and the feedback to ensure stability of the system. Parameters β, α can be estimated online or offline.

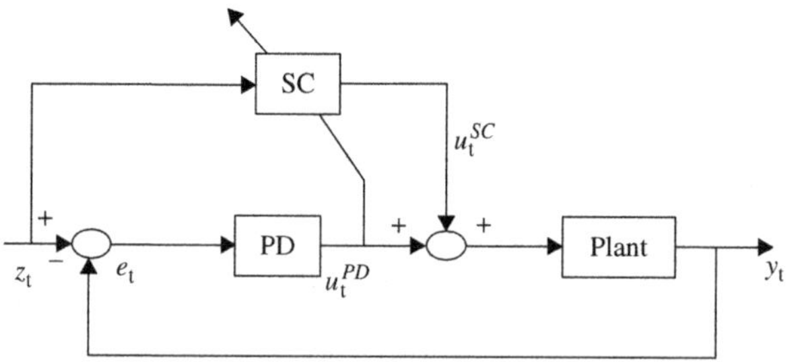

Figure 1.1 Functional scheme of the supervisory control system.

1.2 THE RECURSIVE LEAST SQUARES ALGORITHM WITH DIFFUSE AND SOFT INITIALIZATIONS

Assume that the observation model is of the form

$$y_t = C_t\alpha + \xi_t \quad t = 1, 2, \ldots, N, \tag{1.27}$$

where $y_t \in R^1$ is a measured output, $\alpha = (\alpha_1, \alpha_2, \ldots, \alpha_r)^T \in R^r$ is a vector of unknown parameters, $C_t = (c_{1t}, c_{2t}, \ldots, c_{rt}) \in R^{1\times r}$, c_{it}, $i = 1, 2, \ldots, r$ are regressors (inputs), $\xi_t \in R^1$ is a random process that has uncorrelated values, zero expectation and variance $R_t = E[\xi_t^2]$, and N is a sample size.

Regressors in Eq. (1.27) can be specified by deterministic functions or by random. The autoregressive moving average model

$$\begin{aligned} y_t = {} & a_1 y_{t-1} + a_2 y_{t-2} + \ldots + a_m y_{t-m} + b_1 z_{t-1} + b_2 z_{t-2} \\ & + \ldots + b_n z_{t-n} + \xi_t, \quad t = 1, 2, \ldots, N \end{aligned} \tag{1.28}$$

is one of the most important examples of the application of Eq. (1.27) with

$$C_t = (y_{t-1}, y_{t-2}, \ldots, y_{t-m}, z_{t-1}, z_{t-2}, \ldots, z_{t-n}),$$
$$\alpha = (a_1, a_2, \ldots, a_m, b_1, b_2, \ldots, b_n)^T.$$

Suppose that a sample $\{z_t, y_t\}$, $t = 1, 2, \ldots, N$ and the quadratic criterion of quality taking into account available information on Eq. (1.28) until the time t are given

$$J_t(\alpha) = \sum_{k=1}^{t} \lambda^{t-k}(y_k - C_k\alpha)^2, \quad t = 1, 2, \ldots, N, \tag{1.29}$$

where $\lambda \in (0, 1]$ is a parameter that allows to reduce the past observations' influence (forgetting factor).

We seek an estimate of the unknown parameter from the minimum condition Eq. (1.29), using the RLSM which, as is well known, is determined by the relations [37,38]

$$\alpha_t = \alpha_{t-1} + K_t(y_t - C_t\alpha_{t-1}), \tag{1.30}$$

$$P_t = (P_{t-1} - P_{t-1}C_t^T(\lambda + C_t P_{t-1} C_t^T)^{-1} C_t P_{t-1})/\lambda, \tag{1.31}$$

$$K_t = P_t C_t^T, \quad t = 1, 2, \ldots, N, \tag{1.32}$$

where α_t is the optimal estimate of α, derived from all available observations up to the moment t.

The RLSM successfully is applied in a variety of applications related to adaptive filtering, identification of plants, recognizing and adaptive control, due to the high rate of convergence to the optimal solution and the ability to work under varying conditions [38,39]. However, in order to use it you must specify the initial conditions for the vector α_t and the matrix P_t (to initialize algorithm). There are several ways to their specifications [38].

One of the variants consists in determining and use of the standard LSM (nonrecursive) from the initial data block $\{z_t, y_t\}$, $t = 1, 2, \ldots, \overline{N} < N$. Disadvantages of this approach in real-time processing are quite obvious. First of all, there is the need to memorize observations, the appearance of additional matrix operations including matrix inversion, and the need to agree the work of the RLSM with the obtained estimates of α_0 and P_0.

In another approach (soft initialization) we set $\alpha_0 = 0$ and $P_0 = \mu I_r$, where $\mu > 0$ is a parameter selected by the user. The limiting case $\mu \to \infty$ of soft initialization we will call the diffuse initialization of the RLSM.

During the simulation it was found [40,41] that the estimates of the RLSM are subject to strong fluctuations in the initialization period (in transition stage)

$$t \leq tr = \min_t \left\{ t : \sum_{k=1}^{t} C_k^T C_k > 0, \quad t = 1, 2, \ldots, N \right\}$$

for arbitrarily small noise, any regression order and large values $\mu > 0$. In addition, large values μ can lead to the divergence of the algorithm for $t > tr$ [41]. The study of this phenomenon was the subject of several publications [42–46]. In Ref. [42], the theoretical justification of this behavior for autoregressive moving average models in the transition phase is given. In Ref. [43], the conditions under which value

$$\beta_t(\mu) = E[e_t^T(\mu) e_t(\mu)] / ||\alpha||^2$$

does not exceed 1 (conditions for the absence of the overshoot) at the end of the transition phase are derived, where $e_t(\mu) = \alpha_t - \alpha$. In Ref. [46] the behavior $\beta_t(\mu)$ for $\lambda = 1$ and diffuse initialization is studied, using asymptotic expansions of the RLSM characteristics in inverse powers that were obtained in Ref. [45]. They are based on the inversion formula of perturbed matrices that uses the matrix pseudoinverse [47]. Simulation results presented in Refs. [40,41] show also that a relatively small choice of values μ can increase the transition time. Thus, it is unclear how to choose μ.

And finally, the last of the known methods of the initial conditions setting for the RLSM is the accurate initialization algorithm for the regression with scalar output obtained in Refs. [47–49].

The algorithm allows to obtain an estimate of the RLSM coinciding with the estimate of the not recursive least squares.

1.3 DIFFUSE INITIALIZATION OF THE KALMAN FILTER

Consider a linear discrete system of the form

$$x_{t+1} = A_t x_t + B_t w_t, y_t = C_t x_t + D_t \xi_t, t \in T = \{a, a+1, \ldots, N\}, \quad (1.33)$$

where $x_t \in R^n$ is a state vector, $y_t \in R^m$ is a vector of measured outputs, $w_t \in R^r$ and $\xi_t \in R^l$ are uncorrelated random processes with zero expectations and covariance matrixes $E(w_t w_t^T) = I_r, E(\xi_t \xi_t^T) = I_l$, and A_t, B_t, C_t, D_t are known matrices of appropriate dimensions.

It is required to estimate x_t in the absence of statistical information regarding arbitrary components of x_0. In the literature a number of the KF modifications were proposed that are oriented to solution of this problem.

Thus, in Ref. [50] the unknown initial conditions are interpreted as random variables with zero mean and covariance matrix proportional to a large parameter μ and use of the information filter which calculates recursively the inverse of the covariance matrix (the information matrix) of the estimation error is discussed. However, this filter cannot be used in many situations, and the work with the covariance matrix of the estimation error is more preferable than with its inverse [50, p. 1298]. In Ref. [50], it is proposed to interpret the unknown initial conditions as the diffuse random variables with zero mathematical expectations and an arbitrarily large covariance matrix. In Refs. [17,50–52], under certain assumptions about the process model and the diffuse structure of the initial vector, the problem was solved for $\mu \to \infty$. In these works, characteristics of the KF are expressed explicitly as functions of μ and then their limits are found as $\mu \to \infty$ to obtain accurate solutions—the diffuse KF. In Ref. [53], getting the diffuse KF is based on the use of an information filter. In Ref. [54], any assumptions about the statistical nature of the unknown initial conditions are not used. It is noted that this representation is logically justified in contrast to the diffuse initialization when dealing with an infinite covariance matrix of the initial conditions. In this paper a recursive estimation algorithm in the absence of information about all the initial conditions for a stationary model and a nonsingular dynamics matrix, which is a complicated version of the information filter, is proposed. A feature of the algorithm is the ability to obtain an estimate of the state only after performing a certain number of iterations. Its equivalence to algorithm obtained in Ref. [53] in cases of diffuse initial conditions is shown. In Ref. [55], it is proposed to initialize the KF with the help of expectation and covariance matrix of the obtained estimates from an initial sample. However, obtaining such estimates with acceptable accuracy in some cases is problematic (e.g., in the presence of gaps in the observations, estimating parameters of nonlinear systems and nonstationary processes).

As a practical application of filters with diffuse initialization in Refs. [50–52] the problems of econometrics are considered (estimation, prediction, and smoothing of nonstationary time series), in Refs. [53,54] the problem of designing robust to impulse disturbances filters with sliding window and in Ref. [17] the problem of constructing observers restoring the state of a linear system in finite steps number and training NNs.

CHAPTER 2

Diffuse Algorithms for Estimating Parameters of Linear Regression

Contents

2.1 PROBLEM STATEMENT

Consider a linear observation model of the form

$$y_t = C_t\alpha + \xi_t,\ t = 1, 2, ..., N, \tag{2.1}$$

where $y_t = (y_{1t}, y_{2t}, \ldots, y_{mt})^T \in R^m$ is a vector of outputs, $\alpha = (\alpha_1, \alpha_2, \ldots, \alpha_r)^T \in R^r$ is vector of unknown parameters, $C_t = C_t(z_t)$ is an $m \times r$ matrix, $z_t = (z_{1t}, z_{2t}, \ldots, z_{nt})^T \in R^n$ is a vector of inputs, $\xi_t \in R^m$ is a random process which has uncorrelated values, zero expectation, and a covariance matrix R_t, N is a sample size.

Suppose that the sample $\{z_t, y_t\}$, $t = 1, 2, \ldots, N$ and the quadratic criterion of quality taking into account available information on Eq. (2.1) by the time t are given

$$J_t(\alpha) = \sum_{k=1}^{t} \lambda^{t-k}(y_k - C_k\alpha)^T(y_k - C_k\alpha) + \lambda^t(\alpha - \overline{\alpha})^T\overline{P}^{-1}(\alpha - \overline{\alpha})/\mu,$$
$$t = 1, 2, ..., N, \tag{2.2}$$

where $\overline{\alpha} \in R^r$, $\overline{P} \in R^{r \times r}$ $\overline{P} > 0$ are arbitrary vector and matrix, $\mu > 0$, $\lambda \in (0, 1]$ is a forgetting factor.

Diffuse Algorithms for Neural and Neuro-Fuzzy Networks.
DOI: http://dx.doi.org/10.1016/B978-0-12-812609-7.00002-0

A vector of parameters α is found under the condition that $J_t(\alpha)$ is minimal

$$\partial J_t(\alpha)/\partial\alpha = -2\sum_{k=1}^{t}\lambda^{t-k}C_k^T(y_k - C_k\alpha) + 2\lambda^t\overline{P}^{-1}(\alpha - \overline{\alpha})/\mu = 0$$

and the result must be updated after obtaining a new observation. Or, using the notation α_t for the optimal estimate, we contain

$$M_t\alpha_t = q_t,\ t = 1, 2, ..., N, \tag{2.3}$$

where

$$M_t = \sum_{k=1}^{t}\lambda^{t-k}C_k^T C_k + \lambda^t\overline{P}^{-1}/\mu,\ q_t = \sum_{k=1}^{t}\lambda^{t-k}C_k^T y_k + \lambda^t\overline{P}^{-1}\overline{\alpha}/\mu.$$

Let us at first find a recursive representation for α_t. Since

$$\begin{aligned} M_t &= \lambda M_{t-1} + C_t^T C_t,\ M_0 = \overline{P}^{-1}/\mu, \\ q_t &= \lambda q_{t-1} + C_t^T y_t,\ q_0 = \overline{P}^{-1}\overline{\alpha}/\mu,\ t = 1, 2, ..., N, \end{aligned} \tag{2.4}$$

we have

$$\begin{aligned} \alpha_t &= M_t^{-1}q_t = M_t^{-1}(\lambda q_{t-1} + C_t^T y_t) = M_t^{-1}(\lambda M_{t-1}\alpha_{t-1} + C_t^T y_t) \\ &= M_t^{-1}((M_t - C_t^T C_t)\alpha_{t-1} + C_t^T y_t) = \\ &= \alpha_{t-1} + K_t(y_t - C_t\alpha_{t-1}),\ \alpha_0 = \overline{\alpha},\ t = 1, 2, ..., N, \end{aligned} \tag{2.5}$$

where

$$K_t = M_t^{-1}C_t^T = P_t C_t^T. \tag{2.6}$$

In order to find a recursive representation for the matrix $P_t = M_t^{-1}$, we use the following matrix identity

$$A^{-1} = (B^{-1} + CD^{-1}C^T)^{-1} = B - BC(D + C^T BC)^{-1}C^T B, \tag{2.7}$$

for $B = M_{t-1}^{-1}, D = \lambda I_r, C = C_t^T$. This gives

$$\begin{aligned} P_t &= (\lambda P_{t-1}^{-1} + C_t^T C_t)^{-1} \\ &= (P_{t-1} - P_{t-1}C_t^T(\lambda I_m + C_t P_{t-1} C_t^T)^{-1} C_t P_{t-1})/\lambda, \\ P_0 &= \overline{P}\mu,\ t = 1, 2, ..., N. \end{aligned} \tag{2.8}$$

The relations Eqs. (2.4)–(2.6) and (2.8) describe the recursive least squares method (RLSM) with the quality criteria Eq. (2.2) and the soft initialization defined by $\overline{\alpha}$, $\overline{P}$, μ. It is required to study the RLSM properties depending on $\overline{\alpha}$, $\overline{P}$ as $\mu \to \infty$ with the quality criteria Eq. (2.2) for t belonging to a bounded set $T = \{1, 2, \ldots, N\}$.

Suppose now that in the quality criteria instead of forgetting factor, the covariance matrix of the observation noise, which is usually known in engineering applications [56], is used

$$\begin{aligned} J_t(\alpha) = \sum_{k=1}^{t} (y_k - C_k\alpha)^T R_k^{-1} (y_k - C_k\alpha) \\ + (\alpha - \overline{\alpha})^T \overline{P}^{-1} (\alpha - \overline{\alpha})/\mu, \ t = 1, 2, \ldots, N. \end{aligned} \tag{2.9}$$

Performing calculations analogous to the derivation (2.4)–(2.6), (2.8), it is easy to obtain the following relations for a recursive estimation algorithm (the Kalman filter (KF) if you use statistical interpretation of α, $\overline{\alpha}$, $\overline{P}$)

$$\alpha_t = \alpha_{t-1} + K_t(y_t - C_t\alpha_{t-1}), \ \alpha_0 = \overline{\alpha}, \ t = 1, 2, \ldots, N, \tag{2.10}$$

where

$$K_t = M_t^{-1} C_t^T R_t^{-1} = P_t C_t^T R_t^{-1}, \tag{2.11}$$

$$M_t = M_{t-1} + C_t^T R_t^{-1} C_t, \ M_0 = \overline{P}^{-1}/\mu, \tag{2.12}$$

$$P_t = P_{t-1} - P_{t-1} C_t^T (R_t + C_t P_{t-1} C_t^T)^{-1} C_t P_{t-1}, \ P_0 = M_0^{-1} = \overline{P}\mu. \tag{2.13}$$

As in the case of the quality criteria Eq. (2.2) it is required to study the RLSM properties for t belonging to a bounded set $T = \{1, 2, \ldots, N\}$ depending on $\overline{\alpha}$, $\overline{P}$ as $\mu \to \infty$.

In this chapter it is assumed that C_t is a given deterministic matrix function, $T = \{1, 2, \ldots, N\}$ is a bounded set, and μ is a large positive parameter,

$$N \geq t_{tr} = \min_t \left\{ t: \sum_{k=1}^{t} \lambda^{t-k} C_k^T C_k > 0, t = 1, 2, \ldots, N \right\}.$$

Asymptotic expansions of P_t, K_t in inverse powers of μ will be obtained. They are based on the inversion formula of perturbed matrices that use the matrix pseudoinverse. This allows to present in analytical form P_t, K_t as $\mu \to \infty$; to explain the reason for divergences of the RLSM observed in the simulation for a large μ; to offer limit diffuse estimation algorithms as $\mu \to \infty$ that are independent of the large parameter and result in divergence.

2.2 SOFT AND DIFFUSE INITIALIZATIONS

Let us prove two auxiliary results which will be needed further.

Lemma 2.1.
Suppose that Ω_t is an $n \times n$ matrix which is defined by the expression

$$\Omega_t = \Omega_0/\mu + \sum_{s=1}^{t} F_s^T F_s, \; t = 1, 2, \ldots, N, \tag{2.14}$$

where $\Omega_0 > 0$ is an arbitrary $n \times n$ matrix, F_t, $t = 1, 2, \ldots, N$ are arbitrary $m \times n$ matrixes.

Then the matrix Ω_t^{-1} can be expanded in the power series

$$\begin{aligned}\Omega_t^{-1} &= \Omega_0^{-1}(I_n - \Xi_t\Xi_t^{+})\mu + \Xi_t^{+} + \\ &+ \sum_{i=1}^{q}(-1)^i\Omega_0^{-1/2}(\Omega_0^{1/2}\Xi_t^{+}\Omega_0^{1/2})^{i+1}\Omega_0^{-1/2}\mu^{-i} + O(\mu^{-q-1})\end{aligned} \tag{2.15}$$

which converges uniformly in $t \in T = \{1, 2, \ldots, N\}$ for bounded T and sufficiently large values of
μ, where $\Xi_t = \sum_{s=1}^{t} F_s^T F_s$.

Proof
We have

$$\Omega_t = \Omega_0^{1/2}(I_n + \mu\Omega_0^{-1/2}\Xi_t\Omega_0^{-1/2})\Omega_0^{1/2}/\mu.$$

By means of the inversion formula [47] of the perturbed matrices

$$(I_n + \mu V)^{-1} = (I_n - VV^{+}) + V^{+}(V^{+} + \mu I_n)^{-1}, \tag{2.16}$$

where V is arbitrary symmetric matrix and setting $V = \Omega_0^{-1/2}\Xi_t\Omega_0^{-1/2}$, we obtain

$$\begin{aligned}\Omega_t^{-1} &= \Omega_0^{-1/2}(I_n + \mu\Omega_0^{-1/2}\Xi_t\Omega_0^{-1/2})^{-1}\Omega_0^{-1/2}\mu \\ &= \Omega_0^{-1/2}[(I_n - \Omega_0^{-1/2}\Xi_t\Xi_t^{+}\Omega_0^{1/2}) + \Omega_0^{1/2} \\ &\Xi_t^{+}\Omega_0^{1/2}(\Omega_0^{1/2}\Xi_t^{+}\Omega_0^{1/2} + \mu I_n)^{-1}]\Omega_0^{-1/2}\mu.\end{aligned}$$

Let us use the spectral decomposition of the matrices V and V^{+} for the analysis of this expression

$$\Omega_0^{-1/2}\Xi_t\Omega_0^{-1/2} = T_t\Lambda_tT_t^T,\ \Omega_0^{1/2}\Xi_t^+\Omega_0^{1/2} = T_t\Lambda_t^+T_t^T,$$

where $\Lambda_t = diag(\lambda_t(1), \lambda_t(2), ..., \lambda_t(n))$, $\lambda_t(i)$, $t = 1, 2, ..., N$, $i = 1, 2, ..., n$ are eigenvalues of the matrix $\Omega_0^{-1/2}\Xi_t\Omega_0^{-1/2}$, $\Lambda_t^+ = diag(\lambda_t^+(1), \lambda_t^+(2), ..., \lambda_t^+(n))$,

$$\lambda_t^+(i) = \begin{cases} 0, \lambda_t(i) = 0 \\ 1/\lambda_t(i), \lambda_t(i) \neq 0 \end{cases},\ i = 1, 2, ..., n,$$

T_t is a orthogonal matrix, the columns of which are eigenvectors of the matrix $\Omega_0^{-1/2}\Xi_t\Omega_0^{-1/2}$.

If $\mu > 1/\lambda_{\min}$ and $\lambda_t(i) \neq 0$, then the series

$$(\lambda_t^+(i)+\mu)^{-1}\mu = 1 + \sum_{i=1}^{q}(-1)^i[\lambda_t(i)\mu]^{-i} + O(\mu^{-q-1}),\ i = 1, 2, ..., n$$

converges uniformly in $t\in T = \{1, 2, \ldots, N\}$ for bounded T, where

$$\lambda_{\min} = \min\{\lambda_t(i) > 0, t = 1, 2, \ldots, N, i = 1, 2, \ldots, n\}.$$

As $(\lambda_t^+(i)+\mu)^{-1}\mu = 1$ for $\lambda_t(i) = 0$ then the use of matrix notations gives

$$(\Omega_0^{1/2}\Xi_t^+\Omega_0^{1/2}+\mu I_n)^{-1}\mu = T_t(\Lambda_t^+ + \mu I_n)^{-1}T_t^T\mu$$
$$= I_n + \sum_{i=1}^{q}(-1)^i(\Omega_0^{1/2}\Xi_t^+\Omega_0^{1/2})^i\mu^{-i} + O(\mu^{-q-1}), \mu\to\infty.$$

This implies Eq. (2.15).

Lemma 2.2.

For arbitrary $m\times n$ matrices F_t, $t = 1, 2, ..., N$ the following equalities hold

$$(I_n - \Xi_t\Xi_t^+)F_s^T = 0,\ s\leq t,\ t = 1, 2, ..., N, \tag{2.17}$$

where $\Xi_t = \sum_{k=1}^{t} F_k^TF_k$ is an $n\times n$ matrix.

Proof

1. Introduce the notation for an $n\times mt$ matrix $\tilde{F}_t = (F_1^T, F_2^T, \ldots, F_t^T)$. Let the rank of this matrix be equal to k_t and $l_t(1), l_t(2), ..., l_t(k_t)$ are its arbitrary linearly independent columns.

Using the skeletal decomposition [57] gives

$$\tilde{F}_t = L_t \Gamma_t,$$

where

$$L_t = (l_t(1), l_t(2), \ldots, l_t(k_t)),\ \Gamma_t = (\Gamma_t(1), \Gamma_t(2), \ldots, \Gamma_t(t)),$$
$$rank(L_t) = k_t,\ rank(\Gamma_t) = k_t$$

are $n \times k_t$, $k_t \times mt$ matrices, respectively, $\Gamma_t(i)$, and $i = 1, 2, \ldots, t$ are some $k_t \times m$ matrices.

Let us at first show that

$$I_n - \Xi_t \Xi_t^+ = I_n - L_t (L_t^T L_t)^{-1} L_t^T.$$

We have

$$\Xi_t = \tilde{F}_t \tilde{F}_t^T = L_t \tilde{\Gamma}_t L_t^T,$$

where $\tilde{\Gamma}_t = \Gamma_t \Gamma_t^T$. As $\tilde{\Gamma}_t > 0$ (the Gram matrix constructed by linearly independent rows of Γ_t), $rank(L_t) = rank(\tilde{\Gamma}_t L_t^T)$ and L_t is the full rank matrix by columns. Then [47]

$$\Xi_t^+ = (L_t \tilde{\Gamma}_t L_t^T)^+ = (L_t^T)^+ \tilde{\Gamma}_t^{-1} L_t^+,\ L_t^+ = (L_t^T L_t)^{-1} L_t^T.$$

Thus

$$I_n - \Xi_t \Xi_t^+ = I_n - L_t \tilde{\Gamma}_t L_t^T (L_t \tilde{\Gamma}_t L_t^T)^+ = I_n - L_t \tilde{\Gamma}_t L_t^T (L_t^T)^+ \tilde{\Gamma}_t^{-1} L_t^+$$
$$= I_n - L_t \tilde{\Gamma}_t L_t^T L_t (L_t^T L_t)^{-1} \tilde{\Gamma}_t^{-1} (L_t^T L_t)^{-1} L_t^T = I_n - L_t (L_t^T L_t)^{-1} L_t^T.$$

Since $F_s^T = L_t \Gamma_t(s)$, we have

$$(I_n - \Xi_t \Xi_t^+) F_s^T = (I_n - L_t (L_t^T L_t)^{-1} L_t^T) L_t \Gamma_t(s) = 0,\ s = 1, 2, \ldots, t,\ t = 1, 2, \ldots, n.$$

We at first consider the RLSM properties with the forgetting factor.

Theorem 2.1.
1. The matrices P_t and K_t in Eqs. (2.6) and (2.8) can be expanded in the power series.

$$P_t = M_t^{-1} = \lambda^{-t} \overline{P}(I_r - W_t W_t^+)\mu + W_t^+ +$$
$$+ \sum_{i=1}^{q} (-1)^i \lambda^{ti} \overline{P}^{-1/2} (\overline{P}^{1/2} W_t^+ \overline{P}^{1/2})^{i+1} \overline{P}^{-1/2} \mu^{-i} + O(\mu^{-q-1}), \tag{2.18}$$

$$K_t = [W_t^+ + \sum_{i=1}^{q} (-1)^i \lambda^{ti} \overline{P}^{-1/2} (\overline{P}^{1/2} W_t^+ \overline{P}^{1/2})^{i+1} \overline{P}^{-1/2} \mu^{-i}] C_t^T + O(\mu^{-q-1}) \tag{2.19}$$

which converge uniformly in $t \in T = \{1, 2, \ldots, N\}$ for bounded T and sufficiently large values μ, where

$$W_t = \lambda W_{t-1} + C_t^T C_t,\ W_0 = 0_{r \times r}. \tag{2.20}$$

2. Suppose that elements of the vector α are either unknown constants or random quantities whose statistical characteristics are unknown. And in the last case α is not correlated with ξ_t, $t = 1, 2, ..., N$. Then for any $\varepsilon > 0$

$$P(||\alpha_t - \tilde{\alpha}_t|| \geq \varepsilon) = O(\mu^{-q-1}), \quad \mu \to \infty,\ t = 1, 2, ..., N, \tag{2.21}$$

where

$$\tilde{\alpha}_t = \tilde{\alpha}_{t-1} + \tilde{K}_t(y_t - C_t\tilde{\alpha}_{t-1}),\ \tilde{\alpha}_0 = \overline{\alpha}, \tag{2.22}$$

$$\tilde{K}_t = [W_t^+ + \sum_{i=1}^{q} (-1)^i \lambda^{ti} \overline{P}^{-1/2} (\overline{P}^{1/2} W_t^+ \overline{P}^{1/2})^{i+1} \overline{P}^{-1/2} \mu^{-i}] C_t^T. \tag{2.23}$$

Proof

1. As

$$M_t = \sum_{k=1}^{t} \lambda^{t-k} C_k^T C_k + \lambda^t \overline{P}^{-1}/\mu = \lambda^t (\overline{P}^{-1}/\mu + \sum_{k=1}^{t} \lambda^{-k} C_k^T C_k \tag{2.24}$$

then putting in Lemma 2.1.

$$\Omega_t = M_t,\ \Omega_0 = \overline{P}^{-1},\ F_t = \lambda^{-t/2} C_t,$$

we obtain Eq. (2.18). The representation Eq. (2.19) follows from Eq. (2.18), Lemma 2.2. and the equality

$$K_t = M_t^{-1} C_t^T = P_t C_t^T.$$

2. Introducing the notations

$$e_t = \alpha_t - \tilde{\alpha}_t,\ h_t = \alpha_t - \alpha$$

and using Eqs. (2.5) and (2.22), we obtain

$$e_t = (I_r - \tilde{K}_t C_t) e_{t-1} - (K_t - \tilde{K}_t) C_t h_t + (K_t - \tilde{K}_t) \xi_t,\ e_0 = 0,$$
$$h_t = (I_r - \tilde{K}_t C_t) h_{t-1} + \tilde{K}_t \xi_t,\ h_0 = \overline{\alpha} - \alpha.$$

The matrix of second moments of the block vector $x_t = (e_t^T, h_t^T)^T$ satisfies the following matrix difference equation

$$Q_t = A_t Q_{t-1} A_t^T + L_t$$

with initial conditions $Q_0 = 0$ if elements of the vector α are unknown constants and $Q_0 = block\ diag(0_{r\times r}, \overline{Q}_0)$ if these elements are random quantities, where

$$\begin{aligned} A_t &= \begin{pmatrix} I_r - \tilde{K}_t C_t & -(K_t - \tilde{K}_t)C_t \\ 0 & I_r - \tilde{K}_t C_t \end{pmatrix}, \\ L_t &= \begin{pmatrix} (K_t - \tilde{K}_t)R_t(K_t - \tilde{K}_t)^T & (K_t - \tilde{K}_t)R_t\tilde{K}_t^T \\ \tilde{K}_t R_t (K_t - \tilde{K}_t)^T & \tilde{K}_t R_t \tilde{K}_t^T \end{pmatrix}, \\ \overline{Q}_0 &= E[(\overline{\alpha} - \alpha)(\overline{\alpha} - \alpha)^T]. \end{aligned}$$

Since

$$||K_s - \tilde{K}_t|| = O(\mu^{-q-1}),\ t = 1, 2, ..., N,\ \mu \to \infty$$

then this implies

$$E(e_t^T e_t) = O(\mu^{-q-1}),\ t = 1, 2, ..., N,\ \mu \to \infty.$$

Using the Markov's inequality gives for any $\varepsilon > 0$

$$P(||e_t|| \geq \varepsilon) \leq E(||e_t||^2)/\varepsilon^2 = O(\mu^{-q-1}),\ \mu \to \infty,\ t = 1, 2, ..., N.$$

Neglecting in Eqs. (2.18) and (2.19) the terms since the first order of smallness $O(\mu^{-1})$ we get

$$P_t = M_t^{-1} = \lambda^{-t}\overline{P}(I_r - W_t W_t^+)\mu + W_t^+ + O(\mu^{-1}), \qquad (2.25)$$

$$\alpha_t^{dif} = \alpha_{t-1}^{dif} + K_t^{dif}(y_t - C_t\alpha_{t-1}^{dif}),\ \alpha_0^{dif} = \overline{\alpha}, \qquad (2.26)$$

where

$$K_t^{dif} = W_t^+ C_t^T, \qquad (2.27)$$

$$W_t = \lambda W_{t-1} + C_t^T C_t,\ W_0 = 0_{r\times r},\ t = 1, 2, ..., N. \qquad (2.28)$$

The relations Eqs. (2.25)–(2.28) will be called the diffuse estimation algorithm of the linear regression parameter Eq. (2.1).

Keeping in the expansions Eqs. (2.18) and (2.19) the terms of higher order of smallness, it is possible to obtain different estimation algorithms. So keeping terms of the order $O(\mu^{-1})$ gives

$$P_t = M_t^{-1} = \lambda^{-t}\overline{P}(I_r - W_t W_t^{+})\mu + W_t^{+} - \lambda^t (W_t^{+})^2 \overline{P}^{-1}\mu^{-1} + O(\mu^{-2}), \tag{2.29}$$

$$\alpha_t^{\mu} = \alpha_{t-1}^{\mu} + K_t^{\mu}(y_t - C_t \alpha_{t-1}^{\mu}),\ \alpha_0^{\mu} = \overline{\alpha}, \tag{2.30}$$

where

$$K_t^{\mu} = [W_t^{+} - \lambda^t (W_t^{+})^2 \overline{P}^{-1}\mu^{-1}]C_t^T. \tag{2.31}$$

Note that in contrast to the diffuse algorithm, these relations depend on $\overline{P}$.

Consequence 2.1.1.
The diffuse component

$$P_t^{dif} = \lambda^{-t}\overline{P}(I_r - W_t W_t^{+})\mu$$

is the term in expansion P_t which is proportional to a large parameter and it vanishes when $t \geq tr$, where $tr = \min_t\{t: \Pi_t > 0, t = 1, 2, ..., N\}$, $\Pi_t = \sum_{k=1}^{t} C_k^T C_k$.

Indeed, we have

$$\Pi_t = \tilde{C}_t^T \tilde{C}_t,\ W_t = \sum_{k=1}^{t} \lambda^{t-k} C_k^T C_k = \tilde{C}_t^T \Lambda_t \tilde{C}_t,$$

where
$\tilde{C}_t = (C_2^T, C_1^T, ..., C_t^T)^T \in R^{mt \times r}$, $\Lambda_t = block\ diag(\lambda^{t-1} I_m, \lambda^{t-2} I_m, ..., I_m) \in R^{mtmt}$.

As $\det \Lambda_t \neq 0$, then

$$rank(\Pi_t) = rank(W_t) = rank(\tilde{C}_t)$$

and consequently $P_{tr}^{dif} = 0.$

Consequence 2.1.2.
The matrix K_t^{dif} does not depend on the diffuse component as opposed to the matrix P_t and as the function of μ is uniformly bounded in the norm for $t \in T$ as $\mu \to \infty$.

Consequence 2.1.3.
Numerical implementation errors can result in the RLSM divergence for large values of μ. Indeed, let δW_t^{+} be the error connected with calculations of W_t pseudoinverse. Then by Theorem 2.1

$$\begin{aligned} K_t &= M_t^{-1} C_t^T = P_t C_t^T = \\ &= [\lambda^{-t}\overline{P}(I_r - W_t(W_t^+ + \delta W_t^+))\mu + O(1)]C_t^T \\ &= [-\lambda^{-t}\overline{P}W_t\delta W_t^+\mu + O(1)]C_t^T, \ \mu \to \infty. \end{aligned}$$

Thus for $\delta W_t^+ \neq 0$ the matrix K_t becomes dependent on the diffuse component even when $t \geq tr$. As the operation of pseudoinversion, generally speaking, is not continuous, then the resulting change can be substantial and can lead to divergence. Moreover, in this case K_t becomes proportional to a large parameter and so divergence is possible even if the continuity condition

$$rank(W_t) = rank(W_t + \delta W_t)$$

is satisfied and δW_t^+ is arbitrarily small in respect to the norm.

Note that numerically implemented the diffuse algorithm does not have the mentioned distinctive features. This is evidenced by the absence of diffuse components, i.e., quantities proportional to a large parameter in its construction.

Let us illustrate the difference between the properties of the regression parameter estimates in Eq. (2.1) obtained by the RLSM with the soft initialization for large but finite values of μ and the diffuse algorithm using a numerical example.

Example 2.1.
Consider the problem of estimation of the parameter α in the observation model when $m = 1$, $r = 100$, and $R = 0.05^2$. Let the model inputs be generated by two units that are specified by the cascade connection of the gallery operator ('randsvd', 100) from Matlab package with 64-bit grid and the elements of the vector α are normally distributed with mean 0 and standard deviation 10^4.

Fig. 2.1 shows the dependence of the estimation error on t obtained using the RLSM with the soft and diffuse initializations

$$q_t = ||\alpha - \alpha_t||, \ p_t = ||\alpha - \alpha_t^{dif}||,$$

respectively. In both cases the forgetting factor is set equal to 1.

Curves 1, 2, and 3 correspond to the soft initialization with $\mu = 10^6, 10^9, 10^{10}$, respectively, and curve 4 corresponds to the diffuse initializations for $\overline{\alpha} = 0$, $\overline{P} = I_r$. It is seen that for $\mu = 10^6$ you cannot significantly reduce q_t. If $\mu = 10^9$ then the value of q_t decreases, but it remains approximately two times more than p_t when $t > 100$. A further

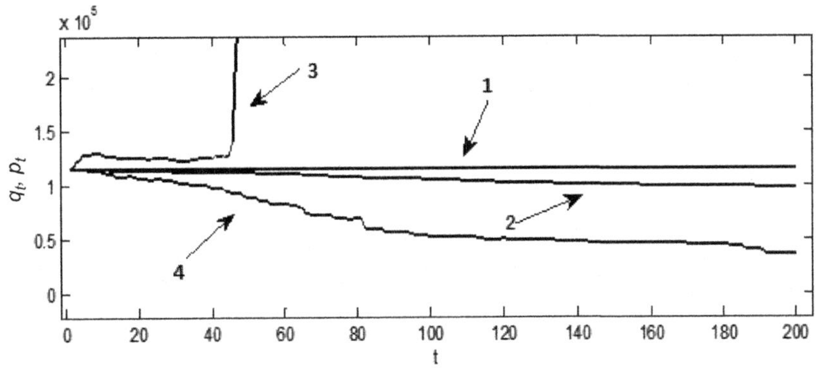

Figure 2.1 Dependence of the estimation error on *t* obtained using the RLSM with the soft and diffuse initializations.

increasing μ leads to divergence of the RLSM that corresponds to Consequence 2.1.3.

Establish some properties of the RLSM with the diffuse initialization. We at first show that the estimates of the least-square method (LSM) coincide with the estimates obtained using the diffusion algorithm from the moment *tr*.

Consider the problem of minimizing the weighted sum of squares

$$\begin{aligned}\tilde{J}_t(\alpha) &= \sum_{k=1}^{t} \lambda^{t-k}(y_k - C_k\alpha)^T(y_k - C_k\alpha) \\ &= \lambda^t(Y_t - \tilde{C}_t\alpha)^T\tilde{R}_t^{-1}(Y_t - \tilde{C}_t\alpha),\ t = 1, 2, ..., N,\end{aligned} \tag{2.32}$$

where

$$\tilde{C}_t = (C_1^T, C_2^T, ..., C_t^T)^T \in R^{mt \times r},\ Y_t = (y_1^T, y_2^T, ..., y_t^T)^T \in R^{mt},$$
$$\tilde{R}_t = block\ diag(\lambda I_m, \lambda^2 I_m, ..., \lambda^t I_m) \in R^{mt \times mt}.$$

The minimum value of α must satisfy the system of equations

$$\partial J_t(\alpha)/\partial\alpha = -2\lambda^t\tilde{C}_t^T\tilde{R}_t^{-1}(Y_t - \tilde{C}_t\alpha) = 0.$$

If the $rank(\tilde{C}_t) = r$ then its solution after t observations will be a linear nonbiased estimate of the form

$$\alpha_t = (\tilde{C}_t^T\tilde{R}_t^{-1}\tilde{C}_t)^{-1}\tilde{C}_t^T\tilde{R}_t^{-1}Y_t. \tag{2.33}$$

We want to show that from the time *tr* the estimates given by the expression Eq. (2.33) and the RLSM with the diffuse initialization coincide. We will need some auxiliary statements.

Lemma 2.3.
For any $t = 2, 3, \ldots$ the following identities are valid

$$(I_r - W_t^+ C_t^T C_t)(I_r - W_{t-1}^+ W_{t-1}) = (I_r - W_t^+ W_t)(I_r - W_{t-1}^+ W_{t-1}), \tag{2.34}$$

$$W_t^+ W_t W_s^+ W_s = W_s^+ W_s, \ s = 1, 2, \ldots, t. \tag{2.35}$$

Proof
Let us prove Eq. (2.34). If $t \geq tr$ then the assertion is obvious. Let $t < tr$. We have

$$I_r - W_t^+ C_t^T C_t = I_r - \lambda^{-t} \Lambda_t^+ C_t^T C_t, \tag{2.36}$$

where Λ_t determined by the matrix equation

$$\Lambda_t = \Lambda_{t-1} + \lambda^{-t} C_t^T C_t, \ \Lambda_0 = 0, \ t = 1, 2, \ldots, N.$$

Transforming the left-hand side of Eq. (2.34) by means of Eq. (2.36), we obtain

$$\begin{aligned}
&(I_r - W_t^+ C_t^T C_t)(I_r - W_{t-1}^+ W_{t-1}) \\
&= (I_r - \lambda^{-t} \Lambda_t^+ C_t^T C_t)(I_r - \Lambda_{t-1}^+ \Lambda_{t-1}) \\
&= I_r - \lambda^{-t} \Lambda_t^+ C_t^T C_t - \Lambda_{t-1}^+ \Lambda_{t-1} + \lambda^{-t} \Lambda_t^+ C_t^T C_t \Lambda_{t-1}^+ \Lambda_{t-1} \\
&= I_r - \Lambda_t^+ \Lambda_t + \Lambda_t^+ \Lambda_{t-1} - \Lambda_{t-1}^+ \Lambda_{t-1} + \Lambda_t^+ \Lambda_t \Lambda_{t-1}^+ \Lambda_{t-1} - \Lambda_t^+ \Lambda_{t-1} \Lambda_{t-1}^+ \Lambda_{t-1} \\
&= (I_r - \Lambda_t^+ \Lambda_t) - (I_r - \Lambda_t^+ \Lambda_t) \Lambda_{t-1}^+ \Lambda_{t-1} + \Lambda_t^+ \Lambda_{t-1} (I_r - \Lambda_{t-1}^+ \Lambda_{t-1}).
\end{aligned}$$

Since pseudoinverse A^+ of any matrix A satisfies the condition $AA^+A = A$ [47], we have

$$\Lambda_t^+ \Lambda_{t-1} (I_r - \Lambda_{t-1}^+ \Lambda_{t-1}) = 0$$

and therefore

$$\begin{aligned}
&(I_r - W_t^+ C_t^T C_t)(I_r - W_{t-1}^+ W_{t-1}) \\
&= (I_r - \Lambda_t^+ \Lambda_t) - (I_r - \Lambda_t^+ \Lambda_t) \Lambda_{t-1}^+ \Lambda_{t-1} \\
&= (I_r - \Lambda_t^+ \Lambda_t)(I_r - \Lambda_{t-1}^+ \Lambda_{t-1}) = (I_r - W_t^+ W_t)(I_r - W_{t-1}^+ W_{t-1}).
\end{aligned}$$

Let us prove Eq. (2.35). If $t \geq tr$ then the assertion is obvious. Suppose now that $t < tr$. Let L_i be a $r \times k_i$ matrix of the rank k_i consisting of all linearly independent columns of the $r \times mi$ matrix

$$\overline{C}_i^T = (\lambda^{-1/2} C_1^T, \lambda^{-1} C_2^T, ..., \lambda^{-i/2} C_i^T)$$

for $i = 1, 2, ..., t$. The matrix L_i is selected so that $L_i = (L_{i-1}, \Delta_i)$, where Δ_i is a $r \times (k_i - k_{i-1})$ matrix of rank $k_i - k_{i-1}$ composed of all linearly independent columns of the matrix $\lambda^{-i/2} C_i^T$ for each $i = 2, 3, ..., t$.

Using the skeletal decomposition yields $\overline{C}_t^T = L_t \Gamma_t$, where $\Gamma_t = (\Gamma_t(1), \Gamma_t(2), ..., \Gamma_t(t))$ is a $k_t \times mt$ matrix of the rank k_t, $\Gamma_t(i)$, $i = 1, 2, ..., t$ are some $k_t \times m$ matrixes, we have

$$\begin{aligned} \Lambda_t &= \sum_{k=1}^{t} \lambda^{-k} C_k^T C_k = \overline{C}_t^T \overline{C}_t = L_t \tilde{\Gamma}_t L_t^T, \\ \Lambda_t^+ &= (L_t \tilde{\Gamma}_t L_t^T)^+ = (L_t^T)^+ \tilde{\Gamma}_t^{-1} L_t^+, \ L_t^+ = (L_t^T L_t)^{-1} L_t^T, \\ W_t^+ W_t &= (L_t \tilde{\Gamma}_t L_t^T)^+ L_t \tilde{\Gamma}_t L_t^T = L_t (L_t^T L_t)^{-1} L_t^T, \end{aligned} \tag{2.37}$$

where $\tilde{\Gamma}_t = \Gamma_t \Gamma_t^T > 0$ is the Gram matrix.

As

$$(L_t^T L_t)^{-1} L_t^T L_s = I_{k_s,k_t}, \ L_t I_{k_s,k_t} = L_s,$$

where $I_{k_s,k_t} = (e_1, e_2, ..., e_s)$, $e_i \in R^{k_t}$ is i-th unit vector then

$$W_t^+ W_t W_s^+ W_s = \Lambda_t^+ \Lambda_t \Lambda_s^+ \Lambda_s = L_t (L_t^T L_t)^{-1} L_t^T L_s (L_s^T L_s)^{-1} L_s^T = W_s^+ W_s.$$

Let us consider the system of equations for the expectation of the estimation error $e_t = E(\alpha - \alpha_t^{dif})$

$$e_t = (I_r - K_t^{dif} C_t) e_{t-1}, \ t = 1, 2, ..., N. \tag{2.38}$$

Establish the form of the normalized transition matrix

$$H_{t,s} = \begin{cases} A_t A_{t-1} ..., A_{s+1}, t = s+1, \ \ s+2, ..., \ \ s \geq 0 \\ I_r, \ \ t = s \end{cases}$$

of this system, where

$$A_t = I_r - K_t^{dif} C_t = I_r - W_t^+ C_t^T C_t.$$

Lemma 2.4.

For any $t = 1, 2, ...$ the following identities are valid

$$\begin{aligned} H_{t,0} &= (I_r - W_t^+ W_t)(I_r - W_{t-1}^+ W_{t-1})...(I_r - W_1^+ W_1) \\ &= I_r - W_t^+ W_t. \end{aligned} \tag{2.39}$$

Proof
We have

$$H_{t,0} = A_t A_{t-1} \ldots A_1, t = 1, 2, \ldots.$$

Let us prove the first equality in Eq. (2.39) using induction. As

$$H_{1,0} = I_r - W_1^+ W_1$$

then it holds at $t = 1$. From Lemma 2.3 it follows that

$$\begin{aligned} H_{2,0} &= (I_r - W_2^+ C_2^T C_2)(I - W_1^+ W_1) \\ &= (I - W_2^+ W_2)(I - W_1^+ W_1) \end{aligned}$$

Thus the first equality in Eq. (2.39) is valid at $t = 2$.

Suppose it is true at some t. Let us show that it will be carried out and at $t + 1$. We have

$$\begin{aligned} H_{t+1,0} &= (I_r - W_{t+1}^+ C_{t+1}^T C_{t+1}) H_{t,0} = \\ &= (I_r - W_{t+1}^+ C_{t+1}^T C_{t+1})(I_r - W_t^+ W_t) q_{t-1}, \end{aligned}$$

where

$$q_{t-1} = (I_r - W_{t-1}^+ W_{t-1}) \ldots (I_r - W_1^+ W_1).$$

Using Lemma 2.3 gives

$$H_{t+1,0} = (I_r - W_{t+1}^+ W_{t+1}) q_t$$

that implies the first equality in Eq. (2.39).

Let us show now that

$$H_{t,0} = I_r - W_t^+ W_t, \quad t = 1, 2, \ldots. \tag{2.40}$$

We use once again induction. As

$$H_{1,0} = I_r - W_1^+ C_1^T C_1 = I_r - W_1^+ W_1$$

then at $t = 1$ Eq. (2.39) is satisfied. Suppose it is true at some t and show that it is true at $t + 1$. Using Lemma 2.3 gives

$$\begin{aligned} H_{t+1,0} &= (I_r - W_{t+1}^+ C_{t+1}^T C_{t+1})(I_r - W_t^+ W_t) = \\ &= (I_r - W_{t+1}^+ W_{t+1})(I_r - W_t^+ W_t) = \\ &= I_r - W_{t+1}^+ W_{t+1} - W_t^+ W_t + W_{t+1}^+ W_{t+1} W_t^+ W_t, = I_r - W_{t+1}^+ W_{t+1}. \end{aligned}$$

Corollary

The solutions of the system Eq. (2.38) satisfy the condition $e_t = 0$ for any initial conditions and $t \geq tr$. That is the parameter estimate obtained by the RLSM with diffuse initialization unbiased when $t \geq tr$.

Note also that in the absence of noise the RLSM with the diffuse initialization restores the vector of unknown parameters in a finite number of steps.

Lemma 2.5.

For any $t \geq tr$

$$H_{t,s} = \lambda^{t-s} W_t^{-1} W_s, \ s = 1, 2, ..., t-1. \tag{2.41}$$

Proof

Let us use induction. We have at $s = t - 1$

$$\begin{aligned} H_{t,t-1} &= A_t = I_r - W_t^{-1} C_t^T C_t \\ &= W_t^{-1}(W_t - C_t^T C_t) = \lambda W_t^{-1} W_{t-1} \end{aligned}$$

Let $s = t - 2$, then

$$\begin{aligned} H_{t,t-2} &= A_t A_{t-1} = \\ &= (I_r - W_t^{-1} C_t^T C_t), \ (I_r - W_{t-1}^{+} C_{t-1}^T C_{t-1}). \end{aligned}$$

Transforming this expression we get

$$\begin{aligned} H_{t,t-2} &= W_t^{-1}(W_t - C_t^T C_t)(I_r - W_{t-1}^{+} C_{t-1}^T C_{t-1}) \\ &= \lambda W_t^{-1} W_{t-1}(I_r - W_{t-1}^{+} C_{t-1}^T C_{t-1}). \end{aligned}$$

Let L_t a matrix of the rank k_t composed of all linearly independent columns of the matrix $\overline{C}_t^T = (\lambda^{(t-1)/2} C_1^T, \lambda^{(t-2)/2} C_2^T, \ldots, C_t^T)$. Using the skeletal decomposition

$$\overline{C}_t^T = L_t \Gamma_t, \ C_t^T = L_t \Gamma_t(t),$$

where L_t and $\Gamma_t = (\Gamma_t(1), \Gamma_t(2), \ldots, \Gamma_t(t))$ are some $mt \times k_t$, $k_t \times r$ matrices of the rank k_t, $\Gamma_t(i)$, $i = 1, 2, \ldots, t$ are some $k_t \times m$ matrices gives

$$W_t W_t^{+} C_t^T = C_t^T. \tag{2.42}$$

It follows from this that

$$H_{t,t-2} = \lambda^2 W_t^{-1} W_{t-2}.$$

Thus Eq. (2.41) is valid at $s = t - 2$.

Let Eq. (2.41) be satisfied at $s = t - i$. We show that it is also valid at $s = t - i - 1$

$$\begin{aligned} H_{t,t-i-1} &= (I_r - W_t^{-1} C_t^T C_t) \ldots (I_r - W_{t-i+1}^{+} C_{t-i+1}^T C_{t-i+1}) \\ &\quad \times (I_r - W_{t-i}^{+} C_{t-i}^T R_{t-i}^{-1} C_{t-i}) = \\ &= \lambda^i W_t^{-1} W_{t-i} (I_r - W_{t-i}^{+} C_{t-i}^T C_{t-i}) = \lambda^{i+1} W_t^{-1} W_{t-i-1}. \end{aligned}$$

Theorem 2.2.

The estimate of the RLSM with diffuse initialization from the moment tr is defined by the expression

$$\alpha_t^{dif} = (\tilde{C}_t^T \tilde{R}_t^{-1} \tilde{C}_t)^{-1} \tilde{C}_t^T \tilde{R}_t^{-1} Y$$

for any initial vector $\overline{\alpha}$, where

$$\begin{aligned} \tilde{C}_t &= (C_1^T, C_2^T, \ldots, C_t^T)^T \in R^{mt \times r}, \; Y_t = (y_1^T, y_2^T, \ldots, y_t^T)^T \in R^{mt}, \\ \tilde{R}_t &= block\ diag(\lambda^{t-1} I_m, \lambda^{t-2} I_m, \ldots, I_m) \in R^{mt \times mt}. \end{aligned}$$

Proof

Iterating Eq. (2.26) and using Lemmas 2.4 and 2.5, we obtain

$$\begin{aligned} \alpha_t^{dif} &= H_{t,0}\overline{\alpha} + \sum_{s=1}^{t} H_{t,s} K_s^{dif} y_s = \\ &= \sum_{s=1}^{t} H_{t,s} K_s^{dif} y_s = W_t^{-1} \sum_{s=1}^{t} \lambda^{t-s} W_s W_s^{+} C_s^T y_s \\ &= W_t^{-1} \sum_{s=1}^{t} \lambda^{t-s} C_s^T y_s = (\tilde{C}_t^T \tilde{R}_t^{-1} \tilde{C}_t)^{-1} \tilde{C}_t^T \tilde{R}_t^{-1} Y_t. \end{aligned}$$

We show now that the estimate of the diffuse algorithm coincides with the LSM with a minimum norm. Let us return again to the problem of minimizing the weighted sum of squares Eq. (2.32). It is known [47] that if rank of the matrix $\tilde{C}_t$ is less than r, then it does not have the unique solution and the solution with minimal norm is given by the expression

$$\alpha_t = \tilde{C}_t^{+} \tilde{R}_t^{-1/2} Y_t, \tag{2.43}$$

where

$$\begin{aligned} \tilde{C}_t &= (C_1^T R_1^{-1/2}, C_2^T R_2^{-1/2}, \ldots, C_t^T R_t^{-1/2})^T \in R^{mt \times r}, Y_t = (y_1^T, y_2^T, \ldots, y_t^T)^T \in R^{mt}, \\ \tilde{R}_t &= block\,diag(\lambda I_m, \lambda^2 I_m, \ldots, \lambda^t I_m) \in R^{mt \times mt}, R_i = \lambda^i I_m, i = 1, 2, \ldots, t. \end{aligned}$$

Establish a link between this estimate and the estimate of the RLSM with the diffuse initialization. We need the following auxiliary assertion.

Lemma 2.6.
The following identity holds

$$M_t^+ \tilde{C}_t^T = [(I_r - K_t C_t) M_{t-1}^+ \tilde{C}_{t-1}^T, K_t R_t^{1/2}],\ t = 1, 2, \ldots, N, \tag{2.44}$$

where $M_t = \tilde{C}_t^T \tilde{C}_t$, $K_t = M_t^+ C_t^T R_t^{-1}$.

Proof
Since

$$M_t^+ \tilde{C}_t^T = M_t^+ (\tilde{C}_{t-1}^T, C_t^T R_t^{-1/2})^T$$

then the last m columns on the left- and right-hand sides of Eq. (2.44) are the same. Let us show that

$$M_t^+ \tilde{C}_{t-1}^T = (I_r - K_t C_t) M_{t-1}^+ \tilde{C}_{t-1}^T.$$

Using the recursive formula

$$M_t = M_{t-1} + \lambda^{-t} C_t^T C_t,\ \Lambda_0 = 0,\ t = 1, 2, \ldots, N,$$

we represent this expression in the equivalent form

$$\begin{aligned}
&(M_t^+ - (I_r - \lambda^{-t/2} K_t C_t) M_{t-1}^+) \tilde{C}_{t-1}^T \\
&= [M_t^+ - (I_r - M_t^+ M_t + M_t^+ M_{t-1}) M_{t-1}^+] \tilde{C}_{t-1}^T \\
&= [M_t^+ (I_r - M_{t-1} M_{t-1}^+) - (I_r - M_t^+ M_t) M_{t-1}^+] \tilde{C}_{t-1}^T = 0.
\end{aligned} \tag{2.45}$$

We show at first that

$$(I_r - M_t^+ M_t) M_{t-1}^+ = 0. \tag{2.46}$$

Suppose that the matrix $\tilde{C}_t$ has a rank k_t and its $l_t(1), l_t(2), \ldots, l_t(k_t)$ linearly independent columns are selected in the same way as the columns of the matrix $\overline{C}_t$ used in the proof of Lemma 2.3. The linear space $C(l_t(1), l_t(2), \ldots, l_t(k_t))$ defined by them coincides with the space formed by the columns M_t. It follows from this that $C(M_{t-1}) \subseteq C(M_t)$.

Using the skeleton decomposition of the matrices, we get

$$\tilde{C}_t^T = L_t \Gamma_t,$$

where $L_t = (l_t(1), l_t(2), \ldots, l_t(k_t))$, $\Gamma_t = (\Gamma_t(1), \Gamma_t(2), \ldots, \Gamma_t(t))$, $rank(\Gamma_t) = k_t$. This implies $M_t = \tilde{C}_t^T \tilde{C}_t = L_t \tilde{\Gamma}_t L_t^T$, where $\tilde{\Gamma}_t = \Gamma_t \Gamma_t^T$.

As the matrix L_t is of full rank by column then $L_t^+ = (L_t^T L_t)^{-1} L_t^T$ and

$$M_t^+ = (L_t \tilde{\Gamma}_t L_t^T)^+ = (L_t^T)^+ \tilde{\Gamma}_t^{-1} (L_t)^+, \; M_t^+ M_t = L_t (L_t^T L_t)^{-1} L_t^T.$$

Substituting these expressions in Eq. (2.46) gives

$$(I_r - L_t (L_t^T L_t)^{-1} L_t^T) L_{t-1} (L_{t-1}^T L_{t-1})^{-1} \tilde{\Gamma}_{t-1}^{-1} L_{t-1}^+ = 0.$$

But since

$$(I_r - L_t (L_t^T L_t)^{-1} L_t^T) L_{t-1} = 0,$$

then Eq. (2.46) is actually performed.

The identity

$$M_t^+ (I_r - M_{t-1} M_{t-1}^+) \tilde{C}_{t-1}^T = 0 \tag{2.47}$$

follows from the equality

$$(I_r - L_{t-1} (L_{t-1}^T L_{t-1})^{-1} L_{t-1}^T) L_{t-1} \Gamma(t-1) = 0.$$

Theorem 2.3.

For any $t = 1, 2, \ldots$ the solution to the optimization problem Eq. (2.32) given by Eq. (2.43) can be obtained by the RLSM with the diffuse initialization and the initial condition $\alpha_0 = 0$.

Proof

Let us show by induction that

$$\alpha_t = \alpha_t^{dif}, \; t = 1, 2, \ldots, N. \tag{2.48}$$

If $t = 1$ then the assertion follows from the expressions

$$\begin{aligned} \alpha_1^{dif} &= K_1^{dif} y_1 = (C_1^T C_1)^+ C_1^T y_1, \\ \alpha_1 &= \tilde{C}_1^+ \tilde{R}_1^{-1/2} Y_1 = (R_1^{-1/2} C_1)^+ R_1^{-1/2} y_1 = (C_1^T C_1)^+ C_1^T y_1. \end{aligned}$$

Assume that Eq. (2.48) holds at some $t-1$. We show that it will also be carried out at t. Since [47] $\tilde{C}_t^+ = M_t^+ \tilde{C}_t^T$ then the use of Eq. (2.44) gives

$$\begin{aligned} &\alpha_t = \tilde{C}_t^+ \tilde{R}_t^{-1/2} Y_t = [(I_r - K_t^{dif} C_t) M_{t-1}^+ \tilde{C}_{t-1}, K_t^{dif} R_t^{1/2}] \\ &\times (Y_{t-1}^T \tilde{R}_{t-1}^{-1/2}, y_t^T R_t^{-1/2})^T = (I_r - K_t^{dif} C_t) M_{t-1}^+ \tilde{C}_{t-1} \tilde{R}_{t-1}^{-1/2} Y_{t-1} + K_t^{dif} y_t \\ &= (I_r - K_t^{dif} C_t) \alpha_{t-1} + K_t^{dif} y_t = \alpha_{t-1} + K_t^{dif} (y_t - C_t \alpha_{t-1}). \end{aligned}$$

The definition of K_t^{dif} is connected with the need to calculate the pseudoinverse of the matrix high order. We show how to avoid it when $m = 1$ calculating recursively the pseudoinverse of the matrix

$$\tilde{C}_t^T = (\lambda^{(t-1)/2} C_1^T, \lambda^{(t-2)/2} C_2^T, \ldots, C_t^T)^T$$

and using the Greville formula [47]

$$(\tilde{C}_t^+)^T = \begin{pmatrix} (\tilde{C}_{t-1}^+)^T (I_r - C_t^T k_t^T) \\ k_t^T \end{pmatrix}, \tag{2.49}$$

where

$$k_t = \begin{cases} \dfrac{(I_r - \tilde{C}_{t-1}^T (\tilde{C}_{t-1}^+)^T) C_t^T}{\|(I_r - \tilde{C}_{t-1}^T (\tilde{C}_{t-1}^+)^T) C_t^T\|^2}, (I_r - \tilde{C}_{t-1}^T (\tilde{C}_{t-1}^+)^T) C_t^T \neq 0 \\ \dfrac{\tilde{C}_{t-1}^+ (\tilde{C}_{t-1}^+)^T C_t^T}{1 + \|C_{t-1}^+ C_t^T\|^2}, (I_r - \tilde{C}_{t-1}^T (\tilde{C}_{t-1}^+)^T) C_t^T = 0 \end{cases}. \tag{2.50}$$

It follows from Eq. (2.49) that $k_t = \tilde{C}_t^+ e_r = (\tilde{C}_t^T \tilde{C}_t)^+ C_t^T$ and therefore $K_t^{dif} = k_t$.

Let us show that in the transition stage ($t \leq tr$) under the condition

$$rank(\tilde{C}_t) = mt, \ \overline{\alpha} = 0 \tag{2.51}$$

the diffuse algorithm does not depend on the forgetting factor λ.

Initially, we show that

$$H_{t,s} K_s^{dif} = \lambda^{t-s} W_t^+ C_s^T, \ t > s, \tag{2.52}$$

where $H_{t,s}$ is the transition matrix of the system Eq. (2.38). Denote $X_{t,s}$ the normalized transition matrix of the auxiliary system

$$x_t = (I_r - K_t C_t) x_{t-1}, \ t = 1, 2, \ldots, N. \tag{2.53}$$

Establish the form of the matrix $X_{t,s}$. Using Eq. (2.4) we find

$$I_r = M_t^{-1} M_t = \lambda M_t^{-1} M_{t-1} + M_t^{-1} C_t^T C_t.$$

Since

$$I_r - M_t^{-1} C_t^T C_t = \lambda M_t^{-1} M_{t-1}$$

then $X_{t,s}$ satisfies the system

$$X_{t,s} = \lambda M_t^{-1} M_{t-1} X_{t-1,s}.$$

It follows from this that

$$M_tX_{t,s} = \lambda M_{t-1}X_{t-1,s} = \lambda^2 M_{t-2}X_{t-2,s} = \ldots = \lambda^{t-s}M_sX_{s,s},$$

$$X_{t,s} = \lambda^{t-s}M_t^{-1}M_s. \tag{2.54}$$

Using Lemmas 2.1 and 2.2. gives

$$H_{t,s}K_s^{dif} = \lim_{\mu\to\infty}X_{t,s}M_s^{-1}C_s^T = \lambda^{t-s}\lim_{\mu\to\infty}M_t^{-1}C_s^T = \lambda^{t-s}W_t^+C_s^T, \tag{2.55}$$

$$H_{t,0} = \lim_{\mu\to\infty}X_{t,0} = \lambda^t\lim_{\mu\to\infty}M_t^{-1}M_0 = \overline{P}(I_r - W_tW_t^+). \tag{2.56}$$

Since

$$H_{t,0} = \overline{P}(I_r - W_tW_t^+) = \overline{P}^{1/2}(I_r - (\overline{P}^{1/2}W_t)(\overline{P}^{1/2}W_t)^+)\overline{P}^{1/2} =$$

$$= (I_r - W_tW_t^+)\overline{P},$$

then this implies that the system Eq. (2.38) has the normalized transition matrix Eq. (2.39).

Note that proving Lemma 2.4., we did not use the power series expansion Eq. (2.18) which, as an example, for random input generally speaking is not true.

We have for $t \leq tr$

$$\begin{aligned}\alpha_t^{dif} &= H_{t,0}\overline{\alpha} + \sum_{s=1}^{t}H_{t,s}K_s^{dif}y_s = \\ &= (I_r - W_tW_t^+)\overline{\alpha} + W_t^+\sum_{s=1}^{t}\lambda^{t-s}C_s^Ty_s.\end{aligned} \tag{2.57}$$

As

$$\begin{aligned}W_t &= \sum_{s=1}^{t}\lambda^{t-s}C_s^TC_s = \lambda^t\overline{C}_t^T\tilde{R}_t^{-1}\overline{C}_t = \lambda^t\tilde{C}_t^T\tilde{C}_t, \\ W_t^+ &= \lambda^{-t}(\tilde{C}_t^T\tilde{C}_t)^+, \ \sum_{s=1}^{t}\lambda^{t-s}C_s^Ty_s = \lambda^t\tilde{C}_t^T\tilde{R}_t^{-1/2}Y_t, \ \tilde{C}_t = \tilde{R}_t^{-1/2}\overline{C}_t,\end{aligned}$$

where

$$\begin{aligned}\overline{C}_t &= (C_2^T, C_1^T, \ldots, C_t^T)^T,\ \tilde{C}_t = (C_1^T R_1^{-1/2}, C_2^T R_2^{-1/2}, \ldots, C_t^T R_t^{-1/2})^T \\ &= \tilde{R}_t^{-1/2}\overline{C}_t \in R^{mt\times r}, \\ Y_t &= (y_1^T, y_2^T, \ldots, y_t^T)^T \in R^{mt},\ \tilde{R}_t = block\ diag(\lambda I_m, \lambda^2 I_m, \ldots, \lambda^t I_m) \in R^{mt\times mt},\end{aligned}$$

and for any matrix A we have $A^+ = (A^T A)^+ A^T$[47], then the substitution of these expressions into Eq. (2.57) gives

$$\begin{aligned}\alpha_t^{dif} &= (I_r - W_t W_t^+)\overline{\alpha} + W_t^+ \tilde{C}_t^T \tilde{R}_t^{-1/2} Y_t = \\ &= (I_r - W_t W_t^+)\overline{\alpha} + (\tilde{C}_t^T \tilde{C}_t)^+ \tilde{C}_t^T \tilde{R}_t^{-1/2} Y_t = \\ &= (I_r - W_t W_t^+)\overline{\alpha} + \overline{C}_t^+ Y_t.\end{aligned}$$

This implies our assertion.

Let us show that if $rank(\tilde{C}_t) = mt$, then the LSM with a minimum norm Eq. (2.43) does not depend on the forgetting factor λ. Let A, B be arbitrary $n\times p$, $p\times m$ matrices, respectively. It is known [47] that if $rank(A) = rank(B) = p$ then $(AB)^+ = B^+ A^+$. This implies

$$\tilde{C}_t^+ = (\tilde{R}_t^{-1/2}\overline{C}_t)^+ = \overline{C}_t^+ \tilde{R}_t^{1/2}.$$

Substitution of this expression in Eq. (2.43) gives $\alpha_t = \overline{C}_t^+ Y_t$.

To conclude this section, we present analogues of some results obtained for the case when the matrix of the intensities of the noise measurement R_t is used in the quality criteria instead of the forgetting factor. In fact, they are simple consequences of the contained results.

Theorem 2.4.

1. The matrices P_t and K_t in Eqs. (2.11) and (2.13) can be expanded in the power series.

$$\begin{aligned}P_t &= M_t^{-1} = \overline{P}(I_r - W_t W_t^+)\mu + W_t^+ + \\ &+ \sum_{i=1}^{q}(-1)^i \overline{P}^{-1/2}(\overline{P}^{1/2} W_t^+ \overline{P}^{1/2})^{i+1} \overline{P}^{-1/2}\mu^{-i} + O(\mu^{-q-1}),\end{aligned} \tag{2.58}$$

$$K_t = \left[W_t^+ + \sum_{i=1}^{q}(-1)^i \overline{P}^{-1/2}(\overline{P}^{1/2} W_t^+ \overline{P}^{1/2})^{i+1} \overline{P}^{-1/2}\mu^{-i}\right] C_t^T R_t^{-1} + O(\mu^{-q-1}), \tag{2.59}$$

which converge uniformly in $t \in T = \{1, 2, \ldots, N\}$ for bounded T and sufficiently large values of μ, where

$$W_t = W_{t-1} + C_t^T R_t^{-1} C_t, \; W_0 = 0_{r \times r}. \tag{2.60}$$

2. For any $\varepsilon > 0$

$$P(||\alpha_t - \tilde{\alpha}_t|| \geq \varepsilon) = O(\mu^{-q-1}), \quad \mu \to \infty, \; t = 1, 2, \ldots, N, \tag{2.61}$$

where

$$\tilde{\alpha}_t = \tilde{\alpha}_{t-1} + \tilde{K}_t(y_t - C_t \tilde{\alpha}_{t-1}), \; \tilde{\alpha}_0 = \overline{\alpha},$$

$$\tilde{K}_t = [W_t^+ + \sum_{i=1}^{q} (-1)^i \overline{P}^{-1/2} (\overline{P}^{1/2} W_t^+ \overline{P}^{1/2})^{i+1} \overline{P}^{-1/2} \mu^{-i}] C_t^T.$$

Proof

1. As

$$M_t = \overline{P}^{-1}/\mu + \sum_{k=1}^{t} C_k^T R_k^{-1} C_k,$$

then putting in Lemma 2.1.

$$\Omega_t = M_t, \; \Omega_0 = \overline{P}^{-1}, \; F_t = R_t^{-1/2} C_t,$$

we obtain Eq. (2.58). The representation Eq. (2.59) follows from Eq. (2.58), Lemma 2.2., and the equality

$$K_t = M_t^{-1} C_t^T R_t^{-1} = P_t C_t^T R_t^{-1}.$$

2. We omit the proof of Eq. (2.61) which is similar to one given in the derivation of Eq. (2.21) in Theorem 2.1.

Neglecting in Eqs. (2.18) and (2.19) the terms beginning with the first order of smallness $O(\mu^{-1})$ we get

$$\begin{aligned} P_t &= M_t^{-1} = \overline{P}(I_r - W_t W_t^+)\mu + W_t^+ + O(\mu^{-1}), \\ \alpha_t^{dif} &= \alpha_{t-1}^{dif} + K_t^{dif}(y_t - C_t \alpha_{t-1}^{dif}), \; \alpha_0^{dif} = \overline{\alpha}, \end{aligned} \tag{2.62}$$

where

$$K_t^{dif} = W_t^+ C_t^T R_t^{-1}, \; t = 1, 2, \ldots, N. \tag{2.63}$$

Note that analogues of the Consequences 2.1.1.–2.1.3. are saved in this case. Corollary 2.1.3. can be extended as follows. The expression

$$K_t = [W_t \delta W_t^+ \mu + O(1)] C_t^T R_t^{-1}, \ \mu \rightarrow \infty$$

implies that the effect of divergence may be increased by high precision measurements (small values of $||R_t||$).

Note also that under the condition Eq. (2.51), where

$$\tilde{C}_t = (C_1^T R_1^{-1/2}, C_2^T R_2^{-1/2}, \ldots, C_t^T R_t^{-1/2})^T \in R^{mt \times r}$$

the diffuse algorithm does not depend on R_t when $t \leq tr$.

Theorem 2.5.
1. For the transition matrix of the system

$$e_t = (I_r - K_t^{dif} C_t) e_{t-1} = (I_r - W_t^+ C_t^T R_t^{-1} C_t) e_{t-1}, \quad (2.64)$$

the following representations hold

$$H_{t,0} = I_r - W_t^+ W_t, \ t = 1, 2, \ldots, N, \quad (2.65)$$

$$H_{t,s} = W_t^{-1} W_s, \ t \geq tr, \ s = 1, 2, \ldots, t-1. \quad (2.66)$$

2. The estimate α_t^{dif} coincides with the LSM estimate of the minimum norm.

2.3 EXAMPLES OF APPLICATION

2.3.1 Identification of Nonlinear Dynamic Plants

Let us illustrate the use of the RLSM with the diffuse initialization for construction of the functionally connected neural network (NN).

Suppose that a plant is described by a nonlinear model of the autoregressive moving average of the form

$$y_t = \Phi(z_t, \beta)\alpha, + \xi_t, \ t = 1, 2, \ldots, N, \quad (2.67)$$

where $z_t = (y_{t-1}, y_{t-2}, \ldots, y_{t-a}, u_{t-d}, u_{t-d-1}, \ldots, u_{t-d-b}) \in R^{a+b}$ is a vector of inputs, $y_t \in R^1$ is a measured output, $\alpha \in R^r$, $\beta \in R^l$ are vectors of unknown parameters, $\Phi(z_t, \beta)$ is a known function, nonlinear in respect to z_t, β, N is a training set size, $\xi_t \in R^1$ is a random process that has uncorrelated values, zero expectation, and variance $R_t = E[\xi_t^2]$, $a, b, d > 0$ are some integer numbers.

It is required to find estimates β_t, α_t using the input–output pairs $\{z_i, y_i\}$, $i = 1, 2, \ldots, t$.

We use the diffuse algorithms from Section 2.2 and the approach based on the functionally connected NNs. The values β are selected by small random numbers and the vector α is unknown and it is needed to find its estimate using given until t the observations of the input–output pairs. Thus the problem reduces to an estimate of α by the linear observations model

$$y_t = C_t\alpha + \xi_t,\ t = 1, 2, \ldots, N, \tag{2.68}$$

where $C_t = \Phi(z_t, \beta)$.

If we use a perceptron with one hidden layer (1.6) and scalar output, then the elements of the matrix C_t are defined by the expression

$$C_t = (\sigma(a_1 z_t + b_1), \sigma(a_2 z_t + b_2), \ldots, \sigma(a_r z_t + b_r)),$$

where $a_k = (a_{k1}, a_{k2}, \ldots, a_{k(a+b)}),\ k = 1, 2, \ldots, r,\ a_k,\ b_k,\ k = 1, 2, \ldots, r$ are weights and biases, respectively, and $\sigma(\cdot)$ is the activation function (AF).

For the RBNN with one hidden layer (1.9) and scalar output C_t is determined by the expression

$$C_t = (1, \phi(b_1||z_t - a_1||^2), \phi(b_2||z_t - a_2||^2), \ldots, \phi(b_p||z_t - a_r||^2)),$$

where a_k, b_k, $k = 1, 2, \ldots, r$ are centers and scaled factors, respectively, and $\sigma(\cdot)$ is a basis function.

Example 2.2.
Let a plant be described by the linear difference equation [58].

$$y_t = y_{t-1}y_{t-2}(y_{t-1} + 2.5)/(1 + y_{t-1}^2 + y_{t-2}^2) + u_{t-1}. \tag{2.69}$$

A sample of u_t from a uniform distribution on the interval $[-2, 2]$ is used for training and for testing the signal $u_t = \sin(2\pi t/250)$. The plant model is sought in the form

$$y_t = f(y_{t-1}, y_{t-2}, u_{t-1}), \tag{2.70}$$

where $f(\cdot)$ is the multilayer perceptron (1.6) with the sigmoid AF. The output layer weights are estimated only. The weights of the hidden layer and biases are selected from the uniform distribution on the intervals $[-1,\ 1]$ and $[0,\ 1]$, respectively. The size of the training sample is
$N = 2000$, the testing is $N = 500$, $\lambda = 1$, $R_t = 1$.

Figs. 2.2 and 2.3 present outputs of the plant and models with 10 and 5 neurons in the hidden layer (curves 1 and 2), respectively. We show

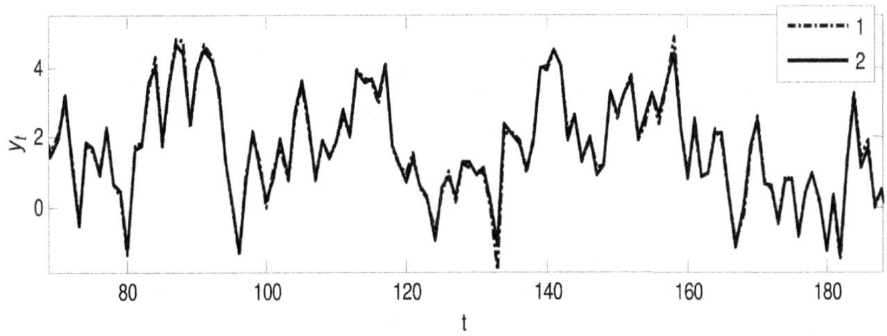

Figure 2.2 Dependencies of the plant output and the model output on time with ten neurons.

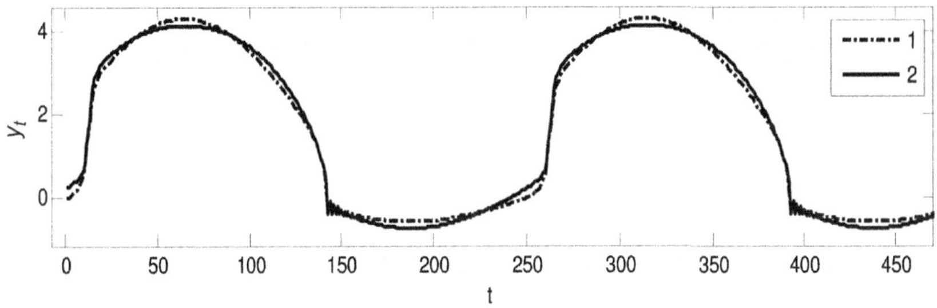

Figure 2.3 Dependencies of the plant output and the model output on time with five neurons.

Table 2.1 Training errors

Sizes of training and testing sets	Number of neurons in the hidden layer	Errors	
		Training set	Testing sets
$N = 2000$	5	0.58	0.6
	10	0.23	0.23
	15	0.14	0.08
	20	0.12	0.044
$N = 500$	4	0.82	0.9
	5	0.57	0.6
	6	0.46	0.46

some realization fragments of testing samples. It can be seen that with 10 neurons in the hidden layer the outputs of the system and its models are visually practically indistinguishable.

Table 2.1 shows the values of the 90th percentile of the mean square error on the training and test sets with $N = 2000$ and $N = 500$, respectively. It is evident that if the number of hidden layer neurons and the training sample size are small, then the identification results with the help of the functionally connected models can be unsatisfactory.

Example 2.3.
Let us use the RBNN and the diffuse algorithm to identify the nonlinear plant described by the equation [59]

$$y_t = u_t^3 + \frac{y_{t-1}}{1 + y_{t-1}^2}. \tag{2.71}$$

As a basic function we choose the Gaussian function, $z_{t1} = u_t$, $z_{t2} = y_t$, $r = 5$,

$$a_1 = (-1, -1),\ a_2 = (-0.5, -0.5),\ a_3 = (0, 0),\ a_4 = (0.5, 0.5),$$
$$b_j = 3.0,\ j = 1, 2, \ldots, 5,\ u_t = \sin(Tt),\ T = 0.001.$$

Figs. 2.4–2.7 show the plant outputs and the diffuse algorithm outputs, the estimation error $e_t = y_t - C_t\alpha_{t-1}^{dif}$ with $\lambda = 0.98$ and $\lambda = 1$, respectively.

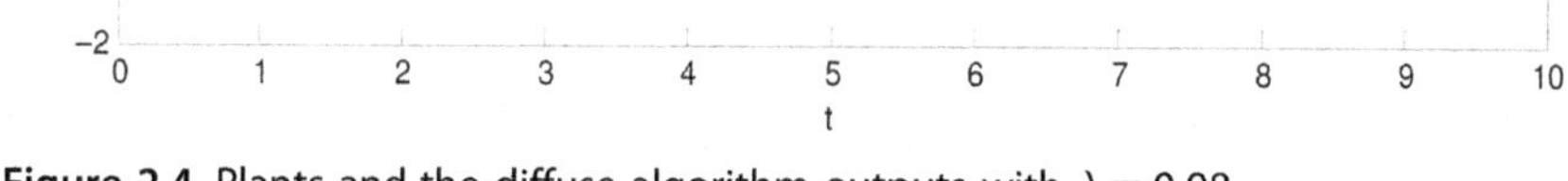

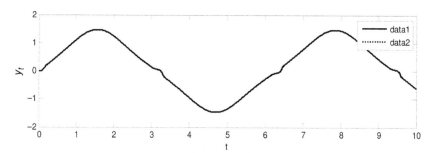

Figure 2.4 Plants and the diffuse algorithm outputs with $\lambda = 0.98$.

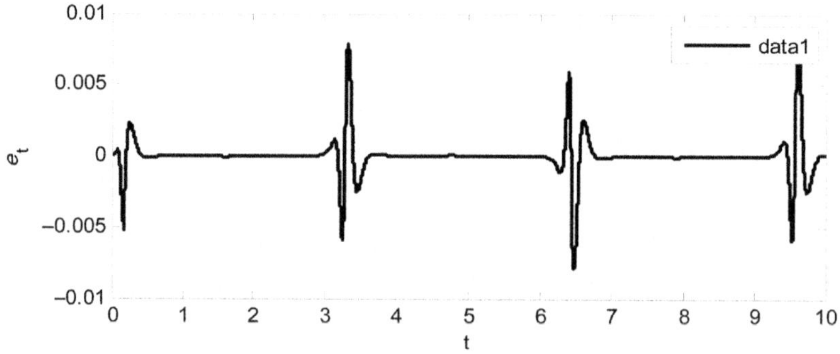

Figure 2.5 The error estimation with $\lambda = 0.98$.

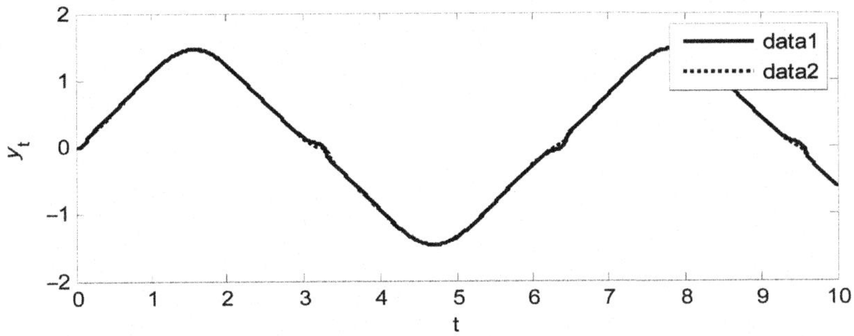

Figure 2.6 Plants and the diffuse algorithm outputs with $\lambda = 1$.

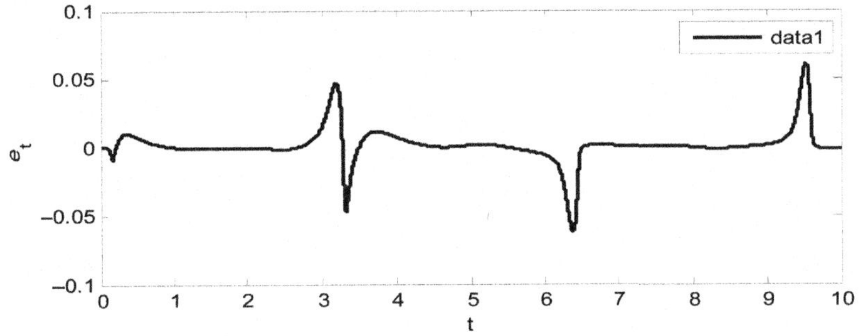

Figure 2.7 The error estimation with $\lambda = 1$.

It can be seen that the use of the forgetting parameter can significantly reduce the error.

2.3.2 Supervisory Control

A functional diagram of the control system is shown in Fig. 1.1.

Suppose that the model of the plant is known and is described by a nonlinear model of the autoregressive moving average

$$y_t = F(z_t, u_t) + \xi_t, \ t = 1, 2, ..., N, \tag{2.72}$$

where $F(\cdot, \cdot)$ is a given nonlinear function, $y_t \in R^1$ is a plant output, $u_t \in R^1$ is a control, $\xi_t \in R^1$ is a random process that has uncorrelated values, zero expectation, and variance $R_t = E[\xi_t^2]$, $z_t = (y_{t-1}, y_{t-2}, ..., y_{t-a})$.

Supervisory control is sought in the form

$$u_t^c = \Phi(r_t, \beta)\alpha, \ t = 1, 2, ..., N, \tag{2.73}$$

where $\alpha \in R^r$, $\beta \in R^l$ are vectors of unknown parameters, r_t is a reference signal, $\Phi(r_t, \beta)$ is a given nonlinear in respect to r_t, β function, and N is a training set size.

PD controller is given by

$$u_t^{pd} = k_p e_t + k_d \dot{e}_t,$$

where $e_t = r_t - y_t$ is a control error, $\dot{e}_t$ is derivative of the control error, k_p, k_d are PD parameters to be selected.

As in the previous section we use an approach based on the functionally connected NN. We assume that the vector β is specified and the vector α is an unknown parameter and to be evaluated. The control problem is formulated as follows. It is required to choose α from the minimum quality criteria

$$J = 1/2 \sum_{i=1}^{t} (u_i^c - r_i)^2 = 1/2 \sum_{i=1}^{t} (\Phi(r_i, \beta)\alpha - r_i)^2$$

for each $t = 1, 2, ..., N$, the PD regulator coefficients from a condition for ensuring the stability of the closed system and the specified quality of the transient process.

We illustrate using a numerical example the behavior of the diffuse estimation algorithm and compare it with the gradient algorithm.

Example 2.4.

Let a plant be linear and described by the transfer function [59]

$$G(p) = \frac{1000}{p^3 + 50p^2 + 2000p}.$$

The functionally connected system is selected in the class of the RBNN with the Gaussian basic function

$$C_t = \Phi(\tilde{r}_t, \beta) = (1, \phi(b_1 || r_t - a_1 ||^2), \phi(b_2 || r_t - a_2 ||^2), ..., \phi(b_5 || r_t - a_5 ||^2)),$$

where $a = (-5, -3, 0, 3, 5)$, $b = (0.5, 0.5, 0\ 5, 0.5)$, $k_p = 30$, $k_d = 0.5$, $T = 0.001$ is the discreteness step, $\alpha = 0.05$ is the learning rate, $\eta = 0.3$ is the impulse coefficient in the gradient algorithm.

Figs. 2.8 and 2.9 present the time dependency of the output of the control plant and the reference signal (curves 1 and 2, respectively) of the supervisory control and the PD controller. The supervisory control values were fixed in the case of the exceeding their absolute value of 10.

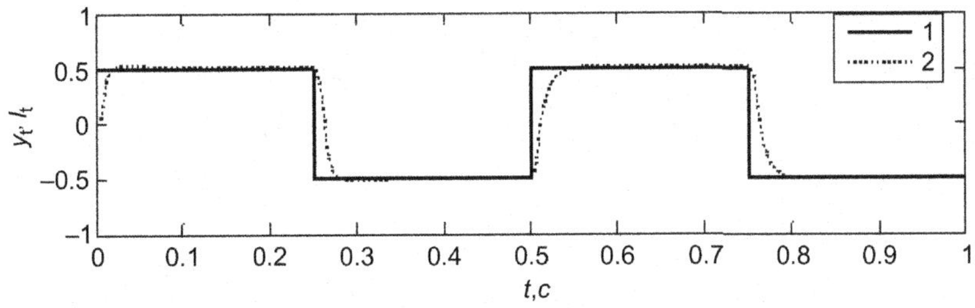

Figure 2.8 Dependencies of the plant output and the reference signal on time.

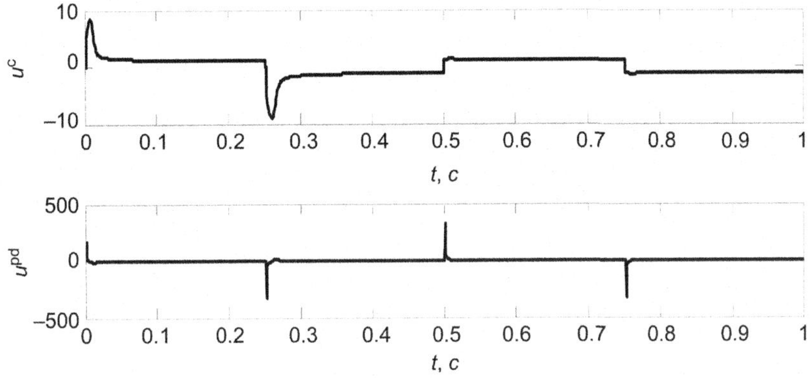

Figure 2.9 Dependencies of the supervisory control and the PD controller on time.

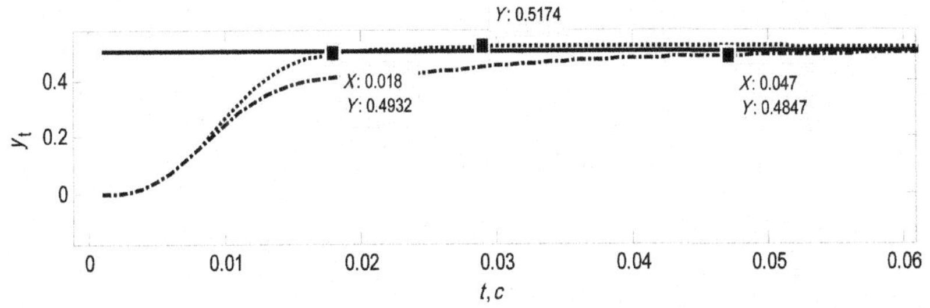

Figure 2.10 Transient processes in a closed system with diffuse algorithm (the *dotted curve*) and the gradient method (the *dash-dotted curve*), the continuous curve is the reference signal r_t.

Fig. 2.10 shows the transition processes in a closed system with diffuse algorithm (the *dashed curve*) and the gradient method (the *bar dotted curve*), the continuous curve is the set point r_t. It is evident that the transition process in the control system with the diffuse supervisory control is substantially shorter than with the gradient algorithm and the overshoot does not exceed 2%.

2.3.3 Estimation With a Sliding Window

One of the known approaches for obtaining robust in respect to perturbations of the estimation algorithms is to use a sliding window [60–64]. In this case the main assumption is based on the adequacy of the system model only on a sliding window interval instead of the entire interval of observations. In these works, preventing divergence of the algorithm and improving the accuracy of estimation are proposed in two ways:

1. Reducing the impact or the rejection of the use of the data outside of the current sliding window.
2. Testing hypotheses about the values of the parameters according to estimates derived from past data and current sliding window and re-initialization, of the estimation algorithm if it is necessary.

In both cases you can use the RLSM with diffuse initialization, which has the property to give unbiased estimates after processing a finite number of observations. If the noise is absent, this algorithm can accurately restore the vector of unknown parameters in a finite number of steps, as opposed to the KF.

Consider the intervals of observations $[t-M, t]$, $t=1,2,...,N$ (sliding windows), where $M > tr$ and suppose that at the moment $t-M$ a priori information about the vector α_{t-M} is absent. Using Theorem 2.4., it is easy to write the following relations to estimate α_t at the moment t based on a sliding window of the latest observations M

$$\alpha_s = \alpha_{s-1} + K_s^{dif}(y_s - C_s\alpha_s),\ \alpha_{t-M} = \overline{\alpha}, \tag{2.74}$$

where

$$K_s^{dif} = W_s^+ C_s^T R_s^{-1}, \tag{2.75}$$

$$\begin{gathered} W_s = W_{s-1} + C_s^T R_s^{-1} C_s,\ W_{t-M} = 0, \\ s = t-M+1, t-M+2, \ldots, t,\ t = M, M+1, \ldots. \end{gathered} \tag{2.76}$$

Example 2.5.

Let us illustrate the effect of using the RLSM with the diffuse initialization in the regime of the sliding window under the parametric uncertainty.

Let there be given the linear model of the autoregressive moving average of the form [64]

$$y_t + a_t y_{t-1} = b_t u_{t-1} + \xi_t,\ t = 1,2,\ldots,N, \tag{2.77}$$

where
$a_t = a - \delta_t,\ b_t = b - \delta_t,\ a = 0.98,\ b = -0.005,\ R = 0.003,$

$$u_t = \sin(0.1t) + 50\cos(200t), \tag{2.78}$$

δ_t is the impulse noise, defined by the expression

$$\delta_t = \begin{cases} 0.05, 200 \leq t \leq 205 \\ 0, \text{othewise} \end{cases}.$$

The simulation results are shown in Figs. 2.11 and 2.12, where $e1(t)$ and $e2(t)$ are the parameter estimation errors of a and b, respectively. Curves 1 and 2 are the estimates obtained with the initialization at $t = 0$ and the estimates obtained using a sliding window in the prediction horizon $M = 50$, respectively. It can be seen that the speed of convergence of the sliding window estimates is significantly higher compared to the speed with the initialization at $t = 0$.

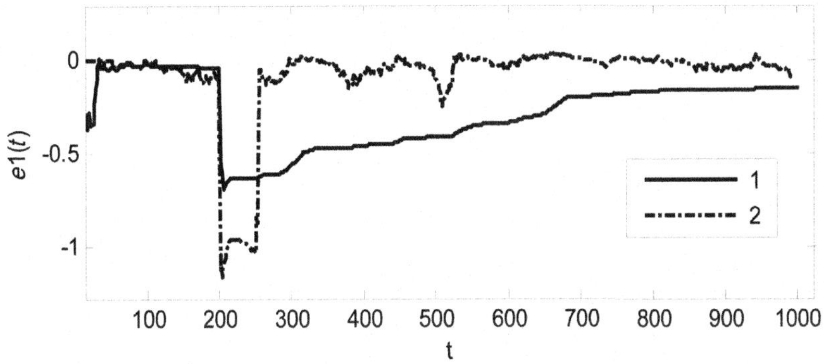

Figure 2.11 Estimation errors of *a*.

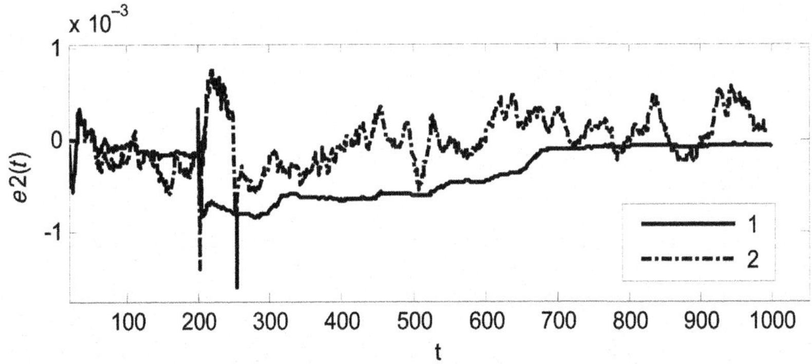

Figure 2.12 Estimation errors of *b*.

Let us show that the RLSM with diffuse initialization can be used together with the assessment procedures of the abrupt change in the parameters of the plant model (change-point). After the abrupt change detection, re-initialization of the algorithm is performed. To illustrate we use the signal model Eq. (2.77) and assume that

$$a_t = \begin{cases} a, t \leq 200 \\ -a, t > 200 \end{cases}, \; b_t = \begin{cases} b, t \leq 200 \\ -b, t > 200 \end{cases}.$$

In this example, changes in the signal model are estimated using a statistical test cumulative sum (CUSUM) [64] according to which the statistics are calculated

$$g_t = g_{t-1} + e_t - \nu, \; g_0 = 0, \; t = 1, 2, ..., N,$$

if $-h < g_t < h$ and otherwise $g_t = 0$, $t_a = t$, where h and ν are the parameters selected by the user, $e_t = y_t - C_t \alpha_t$. The simulation results are shown in Figs. 2.13 and 2.14, where the curves 1 and 2 are the parameter estimates obtained using CUSUM with $h = 100$, $\nu = 0.5$ and the parameter estimates with the initialization at $t = 0$, respectively.

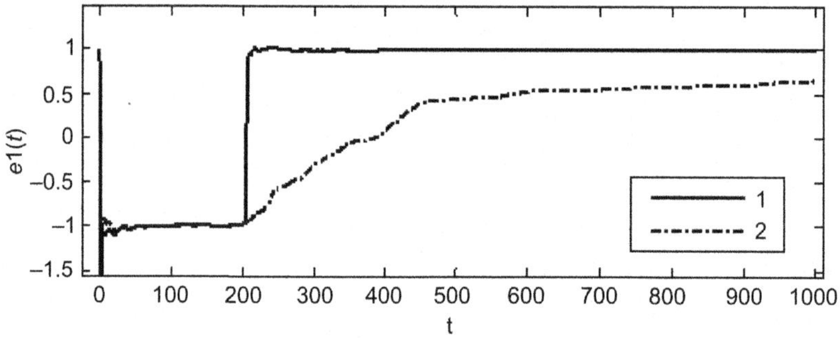

Figure 2.13 Estimation of the parameter *a*.

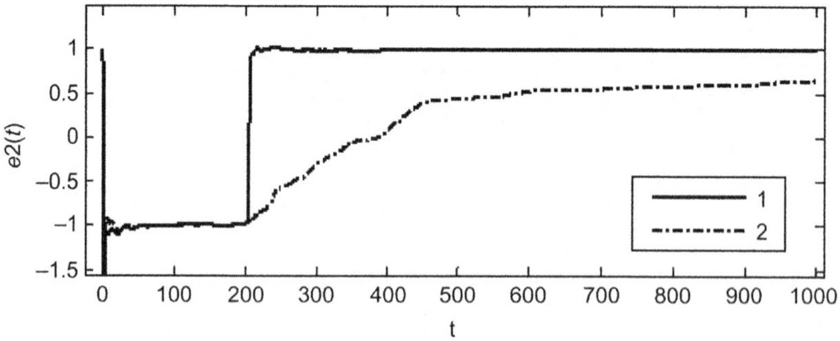

Figure 2.14 Estimation of the parameter *b*.

CHAPTER 3

Statistical Analysis of Fluctuations of Least Squares Algorithm on Finite Time Interval

Contents

3.1 PROBLEM STATEMENT

We restrict ourselves to the most frequently occurring applications of the linear model Eq. (2.1) with the scalar output ($m = 1$) and the constant intensity measurement noise ($R_t = R$). In this case, the recursive least-square method (RLSM) with the forgetting factor and the initializations $\alpha_0 = 0$, $P_0 = I_r\mu$ is determined by the following expressions

$$\alpha_t = \alpha_{t-1} + K_t(y_t - C_t\alpha_{t-1}), \quad \alpha_0 = 0, \tag{3.1}$$

$$K_t = M_t^{-1} C_t^T = P_t C_t^T, \tag{3.2}$$

$$M_t = \lambda M_{t-1} + C_t^T C_t, \quad M_0 = \overline{P}^{-1}/\mu, \tag{3.3}$$

$$P_t = (P_{t-1} - P_{t-1} C_t^T(\lambda + C_t P_{t-1} C_t^T)^{-1} C_t P_{t-1})/\lambda, \ P_0 = I_r\mu, \ \ t = 1, 2, ..., N. \tag{3.4}$$

The components α in Eq. (2.1) are assumed to be unknown constants and t takes the values from a bounded set $t \in T = \{1, 2, \ldots, N\}$, where

$$N > t_{tr} = \min_t \left\{ t : \sum_{k=1}^{t} C_k^T C_k > 0, \quad t = 1, 2, \ldots, N \right\}.$$

Diffuse Algorithms for Neural and Neuro-Fuzzy Networks.
DOI: http://dx.doi.org/10.1016/B978-0-12-812609-7.00003-2

To characterize the statistical properties of the RLSM we will use the bias vector, the matrix of second moments of the estimation error, and a normalized value of the mean square estimation error, respectively

$$a_t(\mu) = E[\alpha_t] - \alpha = E[e_t], \tag{3.5}$$

$$\Psi_t(\mu) = E[e_t e_t^T], \tag{3.6}$$

$$\beta_t(\mu) = E[e_t^T e_t]/||\alpha||^2 = MSE_t(\mu)/||\alpha||^2 = trace[\Psi_t(\mu)]/||\alpha||^2, \tag{3.7}$$

where $e_t = \alpha_t - \alpha$.

As $e_0 = -\alpha$ and $\Psi_0(\mu) = \alpha\alpha^T$ then we have $\beta_0(\mu) = 1$. The function $\beta_t(\mu)$ determines the value of the RLSM overshoot for $t \in T$ and fixed $\mu > 0$: if $\beta_t(\mu) \le 1$, then the overshoot is absent and it is observed if $\beta_t(\mu) > 1$.

It is required to:

1. Study the behavior of $a_t(\mu)$, $\Psi_t(\mu)$ and $\beta_t(\mu)$ for $t \in T$ and $\mu > 0$.
2. Obtain conditions of the RLSM overshoot absence with the soft and the diffuse initializations.

3.2 PROPERTIES OF NORMALIZED ROOT MEAN SQUARE ESTIMATION ERROR

In this section we study the behavior of $a_t(\mu)$, $\Psi_t(\mu)$, and $\beta_t(\mu)$ for $t \in T$ and the finite values of $\mu > 0$.

Iterating equations for the estimation error $e_t = \alpha_t - \alpha$

$$e_t = (I_r - K_t C_t)e_{t-1} + K_t \xi_t, \quad e_0 = -\alpha, \quad t = 1, 2, \ldots, N. \tag{3.8}$$

gives

$$e_t = -X_{t,0}\alpha + \sum_{s=1}^{t} X_{t,s} K_s \xi_s, \quad t = 1, 2, \ldots, N, \tag{3.9}$$

where $X_{t,s}$ is the transition matrix of the homogeneous system

$$x_t = (I_r - K_t C_t)x_{t-1}, \quad t = 1, 2, \ldots, N. \tag{3.10}$$

Substitution of

$$X_{t,s} = \lambda^{t-s} M_t^{-1} M_s, \; K_s = M_s^{-1} C_s^T$$

in Eq. (3.9) yields

$$e_t = -\lambda^t M_t^{-1}\alpha/\mu + \lambda^t M_t^{-1}\sum_{s=1}^{t}\lambda^{-s}C_s^T\xi_s, \quad t = 1, 2, ..., N. \tag{3.11}$$

Thus

$$a_t(\mu) = E[\alpha_t] - \alpha = -\lambda^t M_t^{-1}\alpha/\mu, \tag{3.12}$$

where

$$M_t = \sum_{k=1}^{t}\lambda^{t-k}C_k^T C_k^T + \lambda^t I_r/\mu, \quad t = 1, 2, ..., N.$$

Using Eqs. (3.11) and (3.12), we obtain

$$\begin{aligned}\Psi_t(\mu) = E[e_t e_t^T] &= \lambda^{2t}M_t^{-1}\alpha\alpha^T M_t^{-1}/\mu^2 + RM_t^{-1}\sum_{k=1}^{t}\lambda^{2(t-k)}C_k^T C_k^T M_t^{-1}\\ &= a_t(\mu)a_t^T(\mu) + \Pi_t(\mu), \quad t = 1, 2, ..., N.\end{aligned} \tag{3.13}$$

It follows from Eqs. (3.12) and (3.13) that the matrix of the second moments of the estimation error $\Psi_t(\mu)$ can be represented as the sum of two components. The first is defined only by the bias and does not depend on the intensity of the measurement noise R. Let us show that the second term in Eq. (3.13) is the covariance matrix of α_t. Introducing the notation

$$\overline{\alpha}_t = \alpha_t - E[\alpha_t]$$

and iterating the system of the equations

$$\overline{\alpha}_t = (I_r - K_t C_t)\overline{\alpha}_{t-1} + K_t\xi_t, \quad \overline{\alpha}_0 = 0, \quad t = 1, 2, \ldots, N,$$

we obtain

$$\overline{\alpha}_t = \lambda^t M_t^{-1}\sum_{s=1}^{t}\lambda^{-s}C_s^T\xi_s, \quad t = 1, 2, ..., N.$$

Therefore, the covariance matrix of the estimate α_t is given by the expression

$$\begin{aligned}\mathrm{cov}(\alpha_t) = E[\overline{\alpha}_t\overline{\alpha}_t^T] &= M_t^{-1}\sum_{k=1}^{t}\sum_{j=1}^{t}\lambda^{2t-k-j}C_k^T E[\xi_k\xi_j^T]C_j^T M_t^{-1}\\ &= RM_t^{-1}\sum_{k=1}^{t}\lambda^{2t-k}C_k^T C_k^T M_t^{-1} = \Pi_t(\mu).\end{aligned}$$

Substitution of Eq. (3.13) into Eq. (3.7) gives

$$\beta_t(\mu) = trace\left[\lambda^{2t} M_t^{-1} \alpha \alpha^T M_t^{-1}/\mu^2 + R M_t^{-1} \sum_{k=1}^{t} \lambda^{2(t-k)} C_k^T C_k^T M_t^{-1}\right]/||\alpha||^2$$
$$= \lambda^{2t} \alpha^T M_t^{-2} \alpha/(\mu^2 ||\alpha||^2) + R trace\left(M_t^{-1} \sum_{k=1}^{t} \lambda^{2(t-k)} C_k^T C_k^T M_t^{-1}\right)/||\alpha||^2$$
$$= \eta_t(\mu) + \gamma_t(\mu). \tag{3.14}$$

It can be seen that the numerator of $\eta_t(\mu)$ is proportional to the square of the bias norm and the numerator of $\gamma_t(\mu)$ equals the sum of the diagonal elements of the covariance matrix of α_t.

Let us continue detailing the expression Eq. (3.7) for the normalized values of the mean square estimation error. We have

$$\tilde{M}_t \leq M_t = \sum_{k=1}^{t} \lambda^{t-k} C_k^T C_k^T + \lambda^t I_r/\mu \leq \overline{M}_t, \tag{3.15}$$

where

$$\tilde{M}_t = \lambda^t \left(\sum_{k=1}^{t} C_k^T C_k^T + I_r/\mu \right) = \lambda^t (\tilde{C}_t^T \tilde{C}_t + I_r/\mu),$$

$$\overline{M}_t = \sum_{k=1}^{t} C_k^T C_k^T + I_r/\mu = \tilde{C}_t^T \tilde{C}_t + I_r/\mu, \quad \tilde{C}_t = (C_1^T, C_2^T, ..., C_t^T)^T.$$

Let A, B, C be positive definite matrices such that $C \leq A \leq B$. Since $B^{-1} \leq A^{-1} \leq C^{-1}$ then

$$\overline{M}_t \leq M_t^{-1} \leq \tilde{M}_t^{-1}.$$

Using the spectral decomposition

$$\tilde{C}_t^T \tilde{C}_t = P_t \Lambda_t P_t^T, \quad P_t^T P_t = I_r,$$

where $\Lambda_t = diag(\lambda_t(1), \lambda_t(2), \ldots, \lambda_t(r))$, $P_t = (p_t(1), p_t(2), \ldots, p_t(r))$ is the matrix whose columns are the eigenvectors of the matrix $\tilde{C}_t^T \tilde{C}_t$ corresponding to its eigenvalues $\lambda_t(i)$, $i = 1, 2, ..., r$, $t = 1, 2, ..., N$, we obtain

$$\overline{M}_t = P_t(\Lambda_t + I_r/\mu) P_t^T, \quad \tilde{M}_t = \lambda^t P_t(\Lambda_t + I_r/\mu) P_t^T,$$

$$\overline{M}_t^{-1} = P_t(\Lambda_t + I_r/\mu)^{-1}P_t^T, \quad \tilde{M}_t^{-1} = \lambda^{-t}P_t(\Lambda_t + I_r/\mu)^{-1}P_t^T.$$

Since P_t is an orthogonal matrix then using the identity

$$trace(AB) = trace(BA),$$

we find

$$\tilde{\eta}_t(\mu) \leq \eta_t(\mu) \leq \overline{\eta}_t(\mu), \tag{3.16}$$

where

$$\begin{aligned}
\overline{\eta}_t(\mu) &= \lambda^{2t} trace(\tilde{M}_t^{-1}\alpha\alpha^T\tilde{M}_t^{-1})/(\mu^2||\alpha||^2) \\
&= trace(P_t(\Lambda_t + I_r/\mu)^{-1}P_t^T\alpha\alpha^T P_t(\Lambda_t + I_r/\mu)^{-1}P_t^T)/(\mu^2||\alpha||^2) \\
&= trace((\Lambda_t + I_r/\mu)^{-1}P_t^T\alpha\alpha^T P_t(\Lambda_t + I_r/\mu)^{-1})/(\mu^2||\alpha||^2) \\
&= \sum_{i=1}^{r}\frac{\tilde{\alpha}_i^2(t)}{(\lambda_t(i) + 1/\mu)^2\mu^2}/||\alpha||^2,
\end{aligned}$$

$$\begin{aligned}
\tilde{\eta}_t(\mu) &= \lambda^{2t} trace(\overline{M}_t^{-1}\alpha\alpha^T\overline{M}_t^{-1})/(\mu^2||\alpha||^2) \\
&= \lambda^{2t} trace(P_t(\Lambda_t + I_r/\mu)^{-1}P_t^T\alpha\alpha^T P_t(\Lambda_t + I_r/\mu)^{-1}P_t^T)/(\mu^2||\alpha||^2) \\
&= \lambda^{2t} trace((\Lambda_t + I_r/\mu)^{-1}P_t^T\alpha\alpha^T P_t(\Lambda_t + I_r/\mu)^{-1})/(\mu^2||\alpha||^2) \\
&= \lambda^{2t}\sum_{i=1}^{r}\frac{\tilde{\alpha}_i^2(t)}{(\lambda_t(i) + 1/\mu)^2\mu^2}/||\alpha||^2,
\end{aligned}$$

$$\tilde{\alpha}(t) = (\tilde{\alpha}_1(t), \tilde{\alpha}_2(t), \ldots, \tilde{\alpha}_r(t))^T = P_t^T\alpha.$$

Suppose that C_t it is not a linear combination of $C_1, C_2, \ldots, C_{t-1}$ for $t = 2, 3, \ldots, tr$. Since in this case

$$rank(\tilde{C}_t^T\tilde{C}_t) = rank\left(\sum_{k=1}^{t} C_k^T C_k\right) = rank(C_1^T, C_2^T, \ldots, C_t^T) = t$$

then $r - t$ of the eigenvalues $\tilde{C}_t^T\tilde{C}_t$ are equal to zero. Assuming without loss of generality that $\lambda_t(i) \neq 0, \quad i = 1, 2, \ldots, t, \quad \lambda_t(i) = 0, \quad i = t + 1, t + 2, \ldots, r$, we obtain

$$\overline{\eta}_t(\mu) = \left[\sum_{i=1}^{t}\frac{\tilde{\alpha}_i^2(t)}{(\lambda_t(i) + 1/\mu)^2\mu^2} + \sum_{i=t+1}^{r}\tilde{\alpha}_i^2(t)\right]/||\alpha||^2, \tag{3.17}$$

$$\tilde{\eta}_t(\mu) = \lambda^{2t}\left[\sum_{i=1}^{t}\frac{\tilde{\alpha}_i^2(t)}{(\lambda_t(i)+1/\mu)^2\mu^2} + \sum_{i=t+1}^{r}\tilde{\alpha}_i^2(t)\right]/||\alpha||^2, \quad t=1,2,...,N. \tag{3.18}$$

Establish analogous expressions for $\gamma_t(\mu)$. We have

$$\lambda^{2t}\overline{M}_t^{-1}W_t\overline{M}_t^{-1} \le M_t^{-1}\sum_{k=1}^{t}\lambda^{2(t-k)}C_k^T C_k^T M_t^{-1} \le \tilde{M}_t^{-1}W_t\tilde{M}_t^{-1},$$

where $W_t = \sum_{k=1}^{t} C_k^T C_k^T$. Whence it follows that

$$\tilde{\gamma}_t(\mu) \le \gamma_t(\mu) \le \overline{\gamma}_t(\mu), \tag{3.19}$$

where

$$\overline{\gamma}_t(\mu) = Rtrace(\tilde{M}_t^{-1}W_t\tilde{M}_t^{-1})/||\alpha||^2, \quad \tilde{\gamma}_t(\mu) = R\lambda^{2t}trace(\overline{M}_t^{-1}W_t\overline{M}_t^{-1})/||\alpha||^2.$$

We have also

$$\tilde{M}_t^{-1}W_t = \lambda^{-t}P_t(\Lambda_t+I_r/\mu)^{-1}P_t^T P_t\Lambda_t P_t^T = \lambda^{-t}P_t(\Lambda_t+I_r/\mu)^{-1}\Lambda_t P_t^T,$$

$$\begin{aligned}\tilde{M}_t^{-1}W_t\tilde{M}_t^{-1} &= \lambda^{-2t}P_t(\Lambda_t+I_r/\mu)^{-1}\Lambda_t P_t^T P_t(\Lambda_t+I_r/\mu)^{-1}P_t^T \\ &= \lambda^{-2t}P_t(\Lambda_t+I_r/\mu)^{-2}\Lambda_t P_t^T = \lambda^{-2t}\sum_{i=1}^{t}\frac{\tilde{\alpha}_i^2(t)}{(\lambda_t(i)+1/\mu)^2}p_i^T(t)p_i(t),\end{aligned}$$

$$\begin{aligned}\overline{M}_t^{-1}W_t\overline{M}_t^{-1} &= P_t(\Lambda_t+I_r/\mu)^{-1}P_t^T P_t\Lambda_t P_t^T P_t(\Lambda_t+I_r/\mu)^{-1}P_t^T \\ &= P_t(\Lambda_t+I_r/\mu)^{-2}\Lambda_t P_t^T = \sum_{i=1}^{t}\frac{\tilde{\alpha}_i^2(t)}{(\lambda_t(i)+1/\mu)^2}p_i^T(t)p_i(t).\end{aligned}$$

Substitution of the obtained expressions for $\tilde{M}_t^{-1}W_t\tilde{M}_t^{-1}$ and $\overline{M}_t^{-1}W_t\overline{M}_t^{-1}$ in Eq. (3.19) and taking into account the orthogonality of columns P_t gives

$$\overline{\gamma}_t(\mu) = Rtrace(\tilde{M}_t^{-1}W_t\tilde{M}_t^{-1})/||\alpha||^2 = R\lambda^{-2t}\sum_{i=1}^{t}\frac{\lambda_t(i)}{(\lambda_t(i)+1/\mu)^2}/||\alpha||^2, \tag{3.20}$$

$$\tilde{\gamma}_t(\mu) = R\lambda^{2t} trace(\overline{M}_t^{-1} W_t \overline{M}_t^{-1})/||\alpha||^2 = R\lambda^{2t} \sum_{i=1}^{t} \frac{\lambda_t(i)}{(\lambda_t(i)+1/\mu)^2}/||\alpha||^2. \tag{3.21}$$

Theorem 3.1.
$\eta_t(\mu)$, $\tilde{\eta}_t(\mu)$, $\overline{\eta}_t(\mu)$ are monotonically decreasing functions and $\gamma_t(\mu)$, $\tilde{\gamma}_t(\mu)$, $\overline{\gamma}_t(\mu)$ are monotonically increasing functions of μ for each fixed $t = 1, 2, ..., N$.

Proof
This assertion follows from the expressions

$$\begin{aligned}\partial\eta_t(\mu)/\partial\mu &= \frac{\partial}{\partial\mu}\left[\lambda^{2t}\alpha^T M_t^{-2}\alpha/(\mu^2||\alpha||^2)\right] \\ &= 2\lambda^{3t}\alpha^T M_t^{-3}\alpha/(\mu^4||\alpha||^2) - 2\lambda^{2t}\alpha^T M_t^{-2}\alpha/(\mu^3||\alpha||^2) \\ &= -2\lambda^{2t}\alpha^T M_t^{-3}\sum_{k=1}^{t}\lambda^{(t-k)}C_k^T C_k^T\alpha/(\mu^3||\alpha||^2) < 0,\end{aligned} \tag{3.22}$$

$$\begin{aligned}\partial\overline{\eta}_t(\mu)/\partial\mu &= 2\sum_{i=1}^{t}\frac{\tilde{\alpha}_i^2(t)}{(\lambda_t(i)+1/\mu)^3\mu^4}/||\alpha||^2 - 2\sum_{i=1}^{t}\frac{\tilde{\alpha}_i^2(t)}{(\lambda_t(i)+1/\mu)^2\mu^3}/||\alpha||^2 \\ &= -2\sum_{i=1}^{t}\frac{\tilde{\alpha}_i^2(t)\lambda_t(i)}{(\lambda_t(i)+1/\mu)^3\mu^3}/||\alpha||^2 < 0,\end{aligned} \tag{3.23}$$

$$\begin{aligned}\partial\tilde{\eta}_t(\mu)/\partial\mu &= 2\lambda^{2t}\sum_{i=1}^{t}\frac{\tilde{\alpha}_i^2(t)}{(\lambda_t(i)+1/\mu)^3\mu^4}/||\alpha||^2 - 2\lambda^{2t}\sum_{i=1}^{t}\frac{\tilde{\alpha}_i^2(t)}{(\lambda_t(i)+1/\mu)^2\mu^3}/||\alpha||^2 \\ &= -2\lambda^{2t}\sum_{i=1}^{t}\frac{\tilde{\alpha}_i^2(t)\lambda_t(i)}{(\lambda_t(i)+1/\mu)^3\mu^3}/||\alpha||^2 < 0,\end{aligned} \tag{3.24}$$

$$\begin{aligned}\partial\gamma_t(\mu)/\partial\mu &= \frac{\partial}{\partial\mu}\left[Rtrace\left(\sum_{k=1}^{t}\lambda^{2(t-k)}C_k^T C_k^T M_t^{-2}\right)/||\alpha||^2\right] \\ &= \left[Rtrace\left(\sum_{k=1}^{t}\lambda^{2(t-k)}C_k^T C_k^T \partial M_t^{-2}/\partial\mu\right)/||\alpha||^2\right] \\ &= Rtrace\left(\sum_{k=1}^{t}\lambda^{2(t-k)}C_k^T C_k^T M_t^{-3}\right)/(\mu^2||\alpha||^2) > 0 > 0,\end{aligned} \tag{3.25}$$

$$\partial\overline{\gamma}_t(\mu)/\partial\mu = 2R\lambda^{-2t}\sum_{i=1}^{t}\frac{\lambda_t(i)}{(\lambda_t(i)+1/\mu)^3}/(\mu^2||\alpha||^2)>0, \tag{3.26}$$

$$\partial\tilde{\gamma}_t(\mu)/\partial\mu = 2R\lambda^{2t}\sum_{i=1}^{t}\frac{\lambda_t(i)}{(\lambda_t(i)+1/\mu)^3}/(\mu^2||\alpha||^2)>0. \tag{3.27}$$

Let us illustrate the obtained results using a numerical example.

Example 3.1.
Consider the problem of estimating the bias and the amplitude of harmonics in a signal of an alternating electric current that is described by the model

$$y_t = A_0 + \sum\nolimits_{i=2j+1, j=0,1,\ldots,4}[A_i \sin(2\pi f i \Delta t) + B_i \cos(2\pi f i \Delta t)] + \xi_t, \quad t = 1,2,\ldots,N$$

where $f = 60$ Hz is a fundamental frequency, $r = 11$, ξ_t is a noise, $A_0 = -0.078$, $A_1 = 2.54$, $B_1 = 4.25$, $A_3 = 2.13$, $B_3 = -0.35$, $A_5 = 0.42$, $B_5 = -1.39$, $A_7 = -0.72$, $B_7 = -0.67$, $A_9 = -0.4$, $B_9 = 0.019$, $\Delta = 1/(fk) - T/k$, k is the number of samples per period $T$of the fundamental frequency. Even harmonics are absent in the model.

Figs. 3.1–3.5 show the simulation results when the signal to noise ratio $S_t = \|C_t\|^2\|\alpha\|^2/R$ is equal to 19.4 dB. Fig. 3.1 shows one of the realizations of the signal, where t is a count number, $k = 40$, $R = 1.5^2$. Figs. 3.2 and 3.3 present dependencies $\beta_t(\mu)$, $\gamma_t(\mu)$, $\eta_t(\mu)$ (curves numbered 1, 2, 3, respectively) on t for $\mu = 12$ and $\mu = 36$, respectively, and Fig. 3.4 on μ for t=11. It is seen that the maximum value of the overshoot (Fig. 3.3) is achieved at the end of the transitional stage at $t = 11$ and its value becomes less than 1 after a few cycles of its completion.

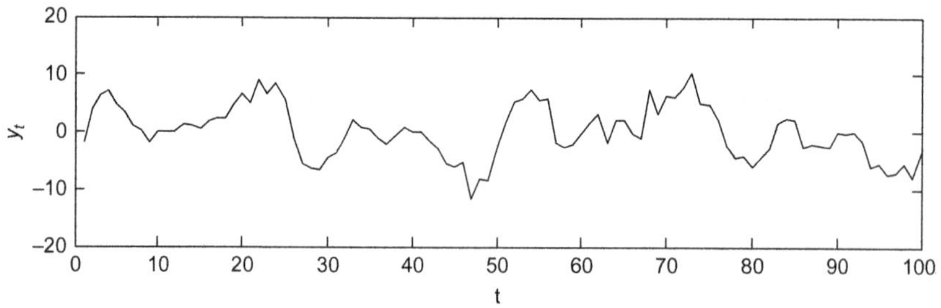

Figure 3.1 Realization of signal.

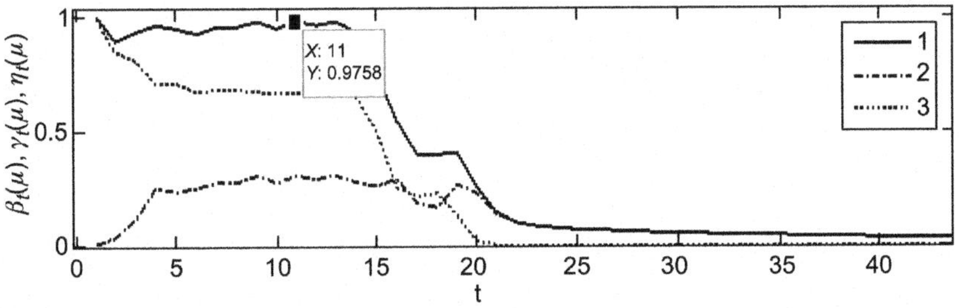

Figure 3.2 Dependencies $\beta_t(\mu)$, $\gamma_t(\mu)$, $\eta_t(\mu)$ on t for $\mu = 12$, $S_t = 19.4$ dB, $k = 40$, $R = 1.5^2$.

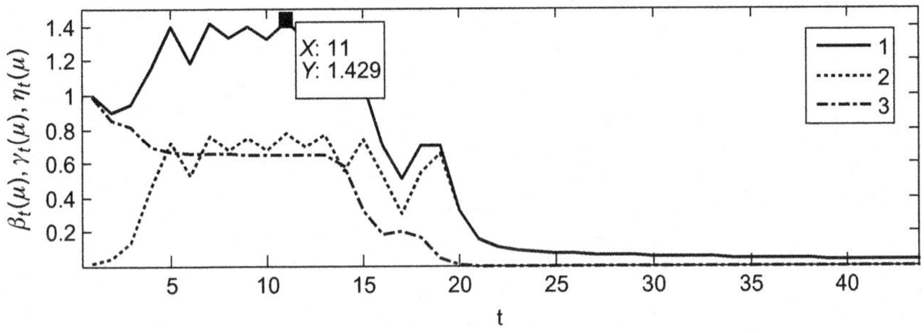

Figure 3.3 Dependencies $\beta_t(\mu)$, $\gamma_t(\mu)$, $\eta_t(\mu)$ on t for $\mu = 36$, $S_t = 19.4$ dB, $k = 40$, $R = 1.5^2$.

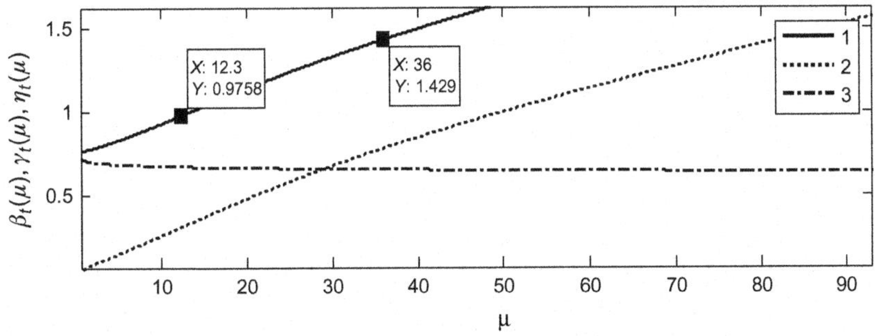

Figure 3.4 Dependencies $\beta_t(\mu)$, $\gamma_t(\mu)$, $\eta_t(\mu)$ on μ, $S_t = 19.4$ dB, $k = 40$, $R = 1.5^2$.

The value $\mu = 12$ corresponds to the absence of overshoot while for $\mu = 36$ it is equal to 1.4.

Furthermore, it is seen that $\gamma_1(\mu)$ is the monotonically increasing function and $\eta_1(\mu)$ is the monotonically decreasing function of μ. Fig. 3.5 shows dependencies $\beta_t(\mu)$ on t for $k = 100$ and $k = 25$ for $\mu = 36$ (curves 1 and 2, respectively). It can be seen that the maximum value of the overshoot is achieved not at the end of the transitional stage $t = 11$ and its duration is substantially longer than the transitional stage duration.

Figs. 3.6–3.8 show similar results with respect to the desired signal to noise ratio in 1 dB and $R = 12.5^2$. It is seen that for no overshoot value μ needs to be significantly reduced compared to the previous case.

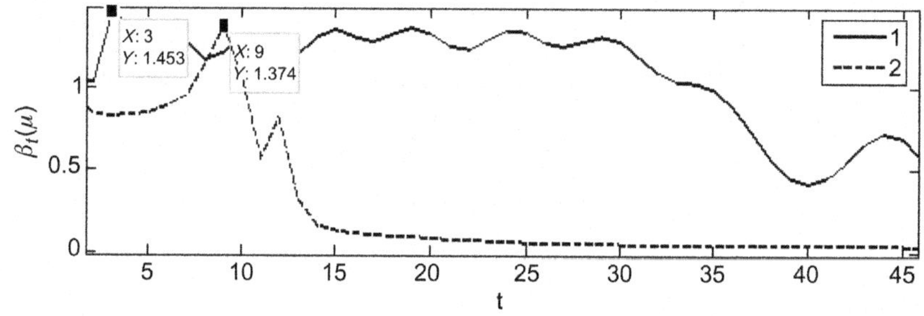

Figure 3.5 Dependencies $\beta_t(\mu)$, $\gamma_t(\mu)$, $\eta_t(\mu)$ on t for $\mu = 36$, $S_t = 19.4$ dB, $k = 100$, $k = 25$, $R = 1.5^2$.

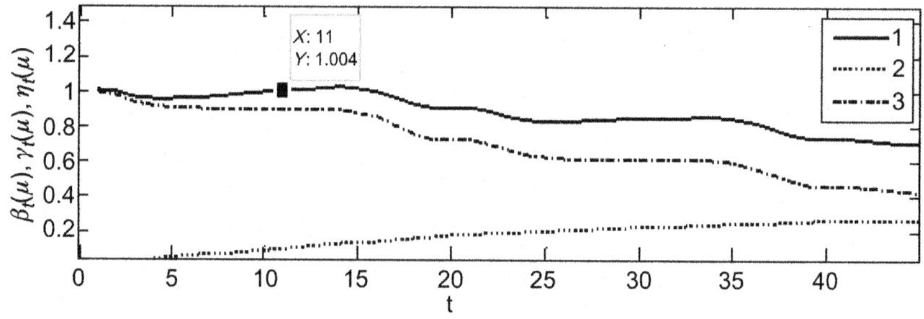

Figure 3.6 Dependencies $\beta_t(\mu)$, $\gamma_t(\mu)$, $\eta_t(\mu)$ on t for $\mu = 0.022$, $S_t = 1$ dB, $k = 40$, $R = 12.5^2$.

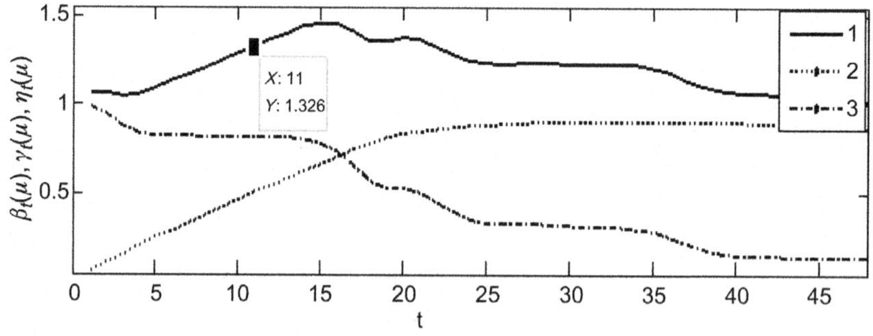

Figure 3.7 Dependencies $\beta_t(\mu)$, $\gamma_t(\mu)$, $\eta_t(\mu)$ on t for $\mu = 0.072$, $S_t = 1$ dB, $k = 40$, $R = 12.5^2$.

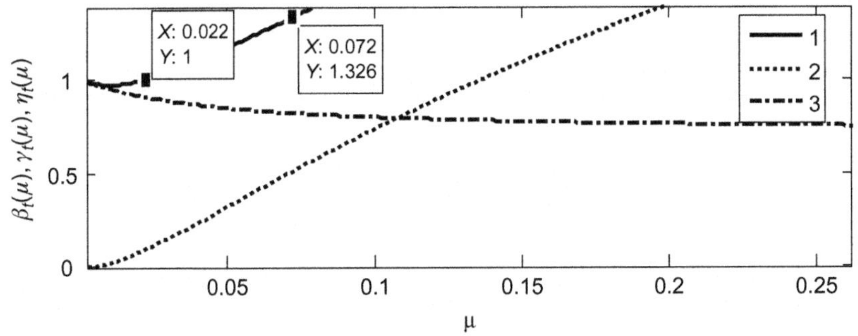

Figure 3.8 Dependencies $\beta_t(\mu)$, $\gamma_t(\mu)$, $\eta_t(\mu)$ on μ, for $S_t = 1$ dB, $k = 40$, $R = 12.5^2$.

Theorem 3.2.

(The existence theorem of overshoot). There are such $\mu_* > 0$, $\mu^* > 0$ that $\mu < \mu_*$ implies

$$\beta_t(\mu) < 1, \quad \tilde{\beta}_t(\mu) = \tilde{\eta}_t(\mu) + \tilde{\gamma}_t(\mu) < 1,$$
$$\overline{\beta}_t(\mu) = \overline{\eta}_t(\mu) + \overline{\gamma}_t(\mu) < 1, t = 1, 2, \ldots, N$$

and $\mu > \mu_*$ implies

$$\beta_t(\mu) > 1, \quad \tilde{\beta}_t(\mu) = \tilde{\eta}_t(\mu) + \tilde{\gamma}_t(\mu) > 1,$$
$$\overline{\beta}_t(\mu) = \overline{\eta}_t(\mu) + \overline{\gamma}_t(\mu) > 1, t = 1, 2, \ldots, N.$$

Proof

Let us prove the statement at first for $\overline{\beta}_t(\mu)$. Since $\overline{\beta}_0(\mu) = 1$ then it is necessary to show that there exists such $\mu_* > 0$ that $\partial\overline{\beta}_t(\mu)/\partial\mu < 0$ for $\mu < \mu_*$ and $\partial\overline{\beta}_t(\mu)/\partial\mu > 0$ for $\mu^* > 0$, $t = 1,2\ldots,tr$.

Using Eqs. (3.17) and (3.20) gives

$$\partial\overline{\beta}_t(\mu)/\partial\mu = \partial\overline{\eta}_t(\mu)/\partial\mu + \partial\overline{\gamma}_t(\mu)/\partial\mu$$
$$= -2\sum_{i=1}^{t} \frac{\tilde{\alpha}_i^2(t)\lambda_i(t)}{(\lambda_i(t)+1/\mu)^3\mu^3}/||\alpha||^2 + 2\lambda^{-2t}R\sum_{i=1}^{t} \frac{\lambda_i(t)}{(\lambda_i(t)+1/\mu)^3}/(\mu^2||\alpha||^2)$$

$$= 2\sum_{i=1}^{t} \frac{\lambda_i(t)}{(\lambda_i(t)+1/\mu)^3}\left[\lambda^{-2t}R - \alpha_i^2(t)/\mu\right]/(\mu^2||\alpha||^2).$$

Whence it follows that $\partial\overline{\beta}_t(\mu)/\partial\mu < 0$ and $\partial\overline{\beta}_t(\mu)/\partial\mu > 0$ if values μ_* and μ^* are defined from the conditions

$$\mu_* < \lambda^{2N}\alpha_{\min}^2/R, \;\; \mu_* > \lambda^2\alpha_{\max}^2/R, \tag{3.28}$$

where $\alpha_{\max}^2 = \max\{\tilde{\alpha}_i^2(t), i = 1, 2, ..., r\}, \alpha_{\min}^2 = \min\{\tilde{\alpha}_i^2(t), i = 1, 2, ..., r\}$, $t = 1, 2, ..., N$.

Using Eqs. (3.18) and (3.21) gives

$$\partial\tilde{\beta}_t(\mu)/\partial\mu = \partial\tilde{\eta}_t(\mu)/\partial\mu + \partial\tilde{\gamma}_t(\mu)/\partial\mu =$$

$$= -2\lambda^{2t}\sum_{i=1}^{t}\frac{\tilde{\alpha}_i^2(t)\lambda_i(t)}{(\lambda_i(t)+1/\mu)^3\mu^3}/||\alpha||^2 + 2R\lambda^{2t}\sum_{i=1}^{t}\frac{\lambda_i(t)}{(\lambda_i(t)+1/\mu)^3}/(\mu^2||\alpha||^2)$$

$$= 2\lambda^{2t}\sum_{i=1}^{t}\frac{\lambda_i(t)}{(\lambda_i(t)+1/\mu)^3\mu^3}\left[R - \tilde{\alpha}_i^2(t)/||\alpha||^2.\right.$$

Whence it follows that the choice of μ using the conditions

$$\mu_* < \alpha_{\min}^2/R, \;\; \mu^* > \alpha_{\max}^2/R \tag{3.29}$$

guarantees in this case the validity of the inequality $\partial\overline{\beta}_t(\mu)/\partial\mu < 0$ for $\mu < \mu_*$ and the inequality $\partial\overline{\beta}_t(\mu)/\partial\mu > 0$ for $\mu > \mu^*$ and any $t = 1,2,...,N$.

Since

$$\tilde{\beta}_t(\mu) \le \beta_t(\mu) \le \overline{\beta}_t(\mu)$$

then for $\mu \le \mu_* = \lambda^{2N}\alpha_{\max}^2/R$ the inequality $\beta_t(\mu) < 1$ is fulfilled and for $\mu \ge \mu_* = \alpha_{\max}^2/R$ the inequality $\beta_t(\mu) > 1$.

The values α and R are unknown and, therefore, the conditions Eqs. (3.28) and (3.29) do not allow you to select the value μ in such a way that there was no overshoot. However, tasks of the signal processing are often accompanied by setting the ratio of signal to noise

$$S_t = \| C_t \|^2 \| \alpha \|^2/R$$

or its lower and upper bounds.

We show now that S_t can be used for the analysis of the overshoot. Let A, B be nonnegative definite matrices. Then, since

$$\lambda_{\min}(A)B \le AB \le \lambda_{\max}(A)B,$$

where $\lambda_{\max}(A)$, $\lambda_{\min}(A)$ are maximum and minimum eigenvalues of the matrix A then taking into account Eq. (3.13), we find

$$\Psi_t(\mu) \ge \lambda^{2t}\alpha\alpha^T/(\lambda_{t,\max}(M_t)\mu)^2 + \Pi_t(\mu),$$

$$\Psi_t(\mu) \le \lambda^{2t}\alpha\alpha^T/(\lambda_{t,\min}(M_t)\mu)^2 + \Pi_t(\mu),$$

where $\lambda_{t,\max}(M_t)$, $\lambda_{t,\min}(M_t)$ are maximum and minimum eigenvalues of the matrix M_t. This implies the estimate

$$\beta_{t,\min}(\mu) \le \beta_t(\mu) \le \beta_{t,\max}(\mu), \quad t = 1, 2, \ldots, N \tag{3.30}$$

where

$$\beta_{tbьum} \ge \lambda^{2t}/(\lambda_{t,\max}(M_t)\mu)^2 + \pi_t(\mu),$$

$$\beta_{t,\max}(\mu) = \lambda^{2t}/(\lambda_{t,\min}(M_t)\mu)^2 + \pi_t(\mu)$$

$$\pi_t(\mu) = \| C_t \|^2 \; trace\left(\sum_{k=1}^{t} \lambda^{2(t-k)} C_k^T C_k M_t^{-2}\right)/S_t \tag{3.31}$$

Example 3.2.

We illustrate the use of Eq. (3.30) in the problem of the harmonics amplitudes estimation in the model of alternating current signal from the previous example.

Figs. 3.9–3.11 show the dependencies $\beta_t(\mu)$, $\beta_{t,\min}(\mu)$, $\beta_{t,\max}(\mu)$ on μ at $t = 11$ for low, medium and large values of the signal to noise ratio $S_t = 1$ dB, $S_t = 10.5$ dB, and $S_t = 21.7$ dB, respectively, obtained from the results of the statistical modeling (curve 2) and calculated by means of Eq. (3.30) (curves 1 and 3, respectively).

Using the obtained results it is easy to evaluate the effect of forgetting factor on the upper and lower limits $\beta_t(\mu)$ in the transitional stage. From Eq. (3.28) it is clear that the introduction of it into a quality criterion could lead to the need for a substantial reduction μ if the order of the model r is quite high. From Eqs. (3.17) and (3.20) it follows that $\overline{\eta}_t(\mu)$

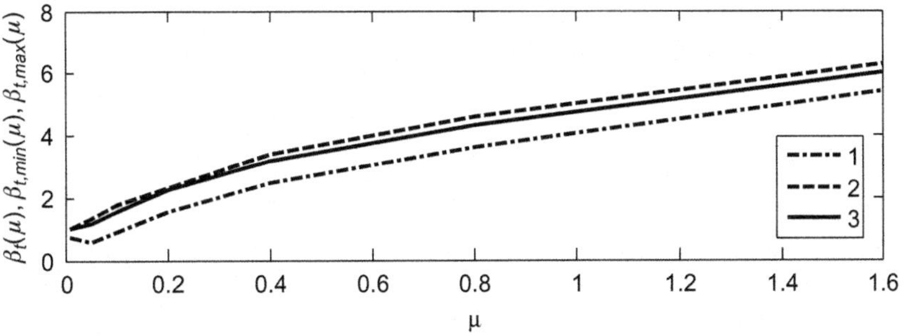

Figure 3.9 Dependencies $\beta_t(\mu)$, $\beta_{t,\min}(\mu)$, $\beta_{t,\max}(\mu)$, $t = 11$, $S_t = 1$ dB.

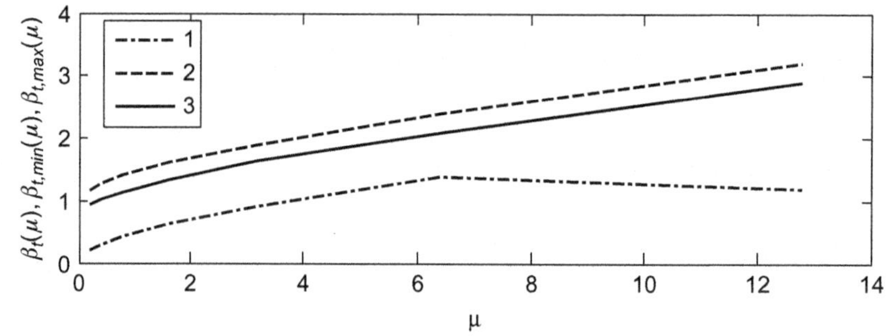

Figure 3.10 Dependencies $\beta_t(\mu)$, $\beta_{t,\min}(\mu)$, $\beta_{t,\max}(\mu)$, $t=11$, $S_t=10.5$ dB

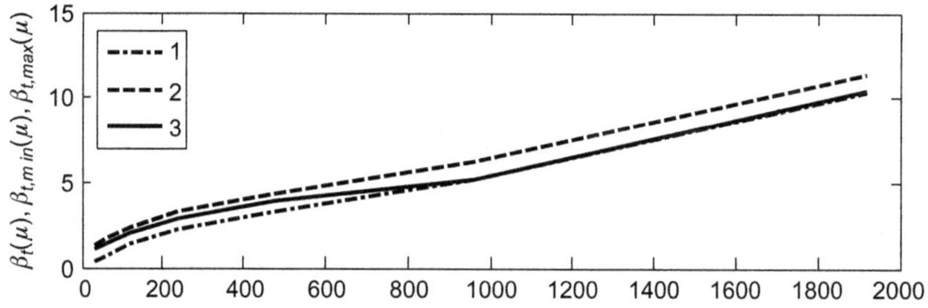

Figure 3.11 Dependencies $\beta_t(\mu)$, $\beta_{t,\min}(\mu)$, $\beta_{t,\max}(\mu)$, $t=11$, $S_t=21.7$ dB

does not depend on λ and $\overline{\gamma}_t(\mu)$ increases for $\lambda<1$ in t. The alternative behavior demonstrates the lower bounds of $\beta_t(\mu)$ decreasing for $\lambda<1$ and increasing in t as it follows from Eqs. (3.18) and (3.21). However, the estimates can be fairly tough. Indeed, in Section 2.2 it was shown that for the diffuse initialization the least squares method (LSM) estimate may not depend on λ. A more subtle result is contained in Theorem 3.3.

Note also that from the above calculations for the case $\lambda=1$ we have

$$\eta_t(\mu)=\left[\sum_{i=1}^{t}\frac{\tilde{\alpha}_i^2(t)}{(\lambda_t(i)+1/\mu)^2\mu^2}+\sum_{i=t+1}^{r}\tilde{\alpha}_i^2(t)\right]/||\alpha||^2, \tag{3.32}$$

$$\gamma_t(\mu)=R\sum_{i=1}^{t}\frac{\lambda_t(i)}{(\lambda_t(i)+1/\mu)^2}/||\alpha||^2. \tag{3.33}$$

Theorem 3.3.

Let us introduce the notations $\eta_t^{1-\varepsilon}(\mu)$, $\gamma_t^{1-\varepsilon}(\mu)$ for $\eta_t(\mu)$, $\gamma_t(\mu)$, respectively, where, $\lambda=1-\varepsilon>0$, $\varepsilon\geq 0$. Then there is ε^* that when $\varepsilon\in(0,\varepsilon^*)$ the following inequalities hold:

1. $\eta_t^{1-\varepsilon}(\mu) \leq \eta_t^1(\mu),\ \gamma_t^{1-\varepsilon}(\mu) \leq \gamma_t^1(\mu), \quad t = 1, 2, ..., tr - 1,$

2. $\eta_t^{1-\varepsilon}(\mu) \leq \eta_t^1(\mu),\ \gamma_t^{1-\varepsilon}(\mu) \leq \gamma_t^1(\mu), \quad t = tr, tr + 1, \ldots, N,$

where $N > t_{tr} = \min_t \left\{ t: \sum_{k=1}^{t} C_k^T C_k > 0, \quad t = 1, 2, \ldots, N \right\}$.

Proof

1. Since, $\eta_t^{1-\varepsilon}(\mu)$, $\gamma_t^{1-\varepsilon}(\mu)$ have continuous partial derivatives with respect to ε for $\varepsilon \in [0, 1)$, $\mu > 0$, $t = 1, 2, ..., N$ then it is sufficient to show that

$$\frac{\partial}{\partial \varepsilon} \eta_t^{1-\varepsilon}(\mu)|_{\varepsilon=0} \leq 0, \quad \frac{\partial}{\partial \varepsilon} \gamma_t^{1-\varepsilon}(\mu)|_{\varepsilon=0} \geq 0.$$

We have for $\eta_t^{1-\varepsilon}(\mu)$:

$$\eta_t^{1-\varepsilon}(\mu) = (1-\varepsilon)^{2t} \alpha^T M_t^{-2} \alpha / (\mu^2 \| \alpha \|^2) = \alpha^T \hat{M}_t^{-2} \alpha / (\mu^2 \| \alpha \|^2),$$

where $\hat{M}_t = \hat{M}_t(\varepsilon) = \sum_{k=1}^{t} (1-\varepsilon)^{-2k} C_k^T C_k + I_r/\mu$.

Since

$$\frac{\partial}{\partial \varepsilon} \hat{M}_t^{-2} = -2\hat{M}_t^{-1} \frac{\partial}{\partial \varepsilon} \hat{M}_t \hat{M}_t^{-1} = -4\hat{M}_t^{-1} \sum_{k=1}^{t} k(1-\varepsilon)^{-2k-1} C_k^T C_k \hat{M}_t^{-1}$$

then

$$\frac{\partial}{\partial \varepsilon} \eta_t^{1-\varepsilon}(\mu)|_{\varepsilon=0} = -4\alpha^T M_t^{-1}(0) \sum_{k=1}^{t} k C_k^T C_k M_t^{-1}(0) \alpha / (\mu^2 \| \alpha \|^2) \leq 0. \tag{3.34}$$

We have for $\gamma_t^{1-\varepsilon}(\mu)$

$$\gamma_t^{1-\varepsilon}(\mu) = Rtrace\left(M_t^{-1} \sum_{k=1}^{t} (1-\varepsilon)^{2(t-k)} C_k^T C_k M_t^{-1} \right) / \| \alpha \|^2$$

$$= Rtrace\left(\hat{M}_t^{-1} \sum_{k=1}^{t} (1-\varepsilon)^{-2k} C_k^T C_k \hat{M}_t^{-1} \right) / \| \alpha \|^2.$$

We have

$$\frac{\partial}{\partial \varepsilon} \left(\hat{M}_t^{-1} \sum_{k=1}^{t} (1-\varepsilon)^{-2k} C_k^T C_k \hat{M}_t^{-1} \right) |_{\varepsilon=0} = I_1 + I_2 + I_3,$$

where

$$
\begin{aligned}
I_1 &= \frac{\partial}{\partial \varepsilon}(\hat{M}_t^{-1})\sum_{k=1}^{t}(1-\varepsilon)^{-2k}C_k^T C_k \hat{M}_t^{-1}|_{\varepsilon=0} \\
&= \hat{M}_t^{-1}(0)\sum_{k=1}^{t} kC_k^T C_k \hat{M}_t^{-1}(0)\sum_{k=1}^{t} C_k^T C_k \hat{M}_t^{-1}(0) \\
&= \hat{M}_t^{-1}(0)\left(\sum_{k=1}^{t} kC_k^T C_k - \sum_{k=1}^{t} kC_k^T C_k \hat{M}_t^{-1}(0)/\mu\right)\hat{M}_t^{-1}(0),
\end{aligned}
$$

$$
I_2 = 2\hat{M}_t^{-1}\sum_{k=1}^{t} k(1-\varepsilon)^{-2k-1}C_k^T C_k \hat{M}_t^{-1}|_{\varepsilon=0} = 2\hat{M}_t^{-1}(0)\sum_{k=1}^{t} kC_k^T C_k \hat{M}_t^{-1}(0),
$$

$$
I_3 = \hat{M}_t^{-1}\sum_{k=1}^{t}(1-\varepsilon)^{-2k}C_k^T C_k \frac{\partial}{\partial \varepsilon}(\hat{M}_t^{-1})|_{\varepsilon=0} = I_1^T,
$$

$$
I_1 + I_2 + I_3 = \hat{M}_t^{-1}(0)\sum_{k=1}^{t} kC_k^T C_k + \sum_{k=1}^{t} kC_k^T C_k \hat{M}_t^{-1}(0)\hat{M}_t^{-1}(0)/\mu.
$$

It follows from this that

$$
\frac{\partial}{\partial \varepsilon}\gamma_t^{1-\varepsilon}(\mu)|_{\varepsilon=0} = 2Rtrace\left(\sum_{k=1}^{t} kC_k^T C_k M_t^{-3}(0)\right)/(\mu\|\alpha\|^2) \geq 0. \qquad (3.35)
$$

2. The statement follows from the facts that $\sum_{k=1}^{t} kC_k^T C_k \geq \sum_{k=1}^{t} C_k^T C_k$ and at $t = tr$ the inequalities Eqs. (3.34) and (3.35) become strict.

Note the differences and the similarities in the formulation and the results of the considered problem and ridge regression [65]. The differences are:

1. The matrix $\sum_{k=1}^{t} \lambda^{(t-k)} C_k^T C_k$ for all $t = 1, 2, \ldots, tr - 1$ is singular and not ill-conditioned.
2. The choice of μ should ensure the condition $\beta_t(\mu) < 1$ for all $t = 1, 2, \ldots, tr$ and not a reduction in the distance between the estimate and the unknown parameter for a fixed value of the sample size.

The similarities are:

1. A priori α is not known so it is unclear how to select μ before the estimation problem solution.

2. The decomposition for the function $\beta_t(\mu)$ describing overshoot into two components proportional to the square of the bias norm and the sum of the diagonal elements of the covariance matrix estimation takes place. The character of their behavior in respect to μ is the same as in the ridge regression.

3.3 FLUCTUATIONS OF ESTIMATES UNDER SOFT INITIALIZATION WITH LARGE PARAMETERS

In this section we will continue the study of the RLSM estimate fluctuations suggesting that μ is a large positive parameter.

At first we obtain the asymptotic formulas derived from Eq. (2.18) and use these for the RLSM estimate fluctuations analysis. It will be convenient to formulate the results in the form of two separate theorems.

Theorem 3.4.
The functions $a_t(\mu)$, $\eta_t(\mu)$, $\prod_t(\mu)$, $\gamma_t(\mu)$ can be expanded in the power series

$$a_t(\mu) = E[e_t] = a_t^0 + a_t^1/\mu + \sum_{i=1}^{\infty}(-1)^{i+1}\lambda^{ti}(W_t^+)^{i+1}\mu^{-i-1}\alpha, \tag{3.36}$$

$$\eta_t(\mu) = \eta_t^0 + \lambda^{2t}\sum_{i=1}^{\infty}(-1)^i(i+1)\lambda^{ti}\alpha^t(W_t^+)^{i+2}\alpha/\|\alpha\|^2\mu^{-i-2}, \tag{3.37}$$

$$\Pi_t(\mu) = \Pi_t^0 + \Pi_t^1/\mu + R\sum_{i=1}^{\infty}(-1)^i(i+3)(W_t^+)^{i+3}\mu^{-i-2}, \quad \lambda = 1, \tag{3.38}$$

$$\gamma_t(\mu) = \gamma_t^0 + \gamma_t^1/\mu + R\sum_{i=1}^{\infty}(-1)^i(i+3)trace((W_t^+)^{i+3})/\|\alpha\|^2\mu^{-i-2}, \quad \lambda = 1 \tag{3.39}$$

which converge uniformly in $t\in T = \{1, 2, \ldots, N\}$ for bounded T and sufficiently large values of μ, where

$$a_t^0 = -(I_n - W_tW_t^+)\alpha, \quad a_t^1 = \lambda^t W_t^+\alpha, \quad \eta_t^0 = \alpha^T(I_r - W_tW_t^+)\alpha/\|\alpha\|^2,$$

$$\Pi_t^0 = RW_t^+/\|\alpha\|^2, \quad \Pi_t^1 = -2R(W_t^+)^2/\|\alpha\|^2, \quad \gamma_t^0 = Rtrace(W_t^+)/\|\alpha\|^2,$$

$$\gamma_t^1 = -2Rtrace(W_t^+)^2/\|\alpha\|^2,$$

$$W_t = \sum_{k=1}^{t} \lambda^{t-k} C_k^T C_k.$$

Proof

The expansion Eq. (3.36) follows from Eqs. (2.12) and (2.18). Let us prove Eq. (3.37). Using Eq. (2.18) gives

$$M_t^{-1} = \lambda^{-t}(I_r - W_t W_t^+)\mu + \sum_{i=0}^{\infty}(-1)^i \lambda^{ti}(W_t^+)^{i+1}\mu^{-i} = I_1 + I_2,$$
$$M_t^{-2} = (I_1 + I_2)^2 = I_1^2 + I_1 I_2 + I_2 I_1 + I_2^2.$$

Simplify the terms in the right-hand side of the expression for M_t^{-2}. We use the following pseudoinversion properties:

1. The matrix $(I_r - W_t W_t^+)$ is idempotent and $W_t^+ W_t W_t^+ = W_t^+$, $W_t W_t^+ W_t = W_t$.
2. For any symmetric matrix A the identity $AA^+ = A^+A$ is valid and therefore

$$W_t^+(I_r - W_t W_t^+) = 0, \quad (I_r - W_t^+ W_t)W_t^+ = (I_r - W_t W_t^+)W_t^+ = 0.$$

It follows from this that

$$I_1^2 = \lambda^{-2t}(I_r - W_t W_t^+)\mu^2, \;\; I_1 I_2 = 0, \;\; I_2 I_1 = 0. \tag{3.40}$$

With the help of the Cauchy formula of two series product presentation [66]

$$\left(\sum_{i=0}^{\infty} a_k\right)\left(\sum_{i=0}^{\infty} b_k\right) = \sum_{n=0}^{\infty}\sum_{i=0}^{n} a_k b_{n-k} \tag{3.41}$$

setting $a_k = b_k = (-1)^k \lambda^{tk}(W_t^+)^{k+1}\mu^{-k}$, we find

$$I_2^2 = \sum_{i=0}^{\infty}(-1)^i(i+1)\lambda^{ti}(W_t^+)^{i+1}\mu^{-i}.$$

From this, taking into account Eq. (3.40), we obtain a uniformly converging in $t = 1, 2, \ldots, N$ and sufficiently large values of μ power series

$$M_t^{-2} = \lambda^{-2t}(I_r - W_t W_t^+)\mu^2 + \sum_{i=0}^{\infty}(-1)^i(i+1)\lambda^{ti}(W_t^+)^{i+2}\mu^{-i}. \tag{3.42}$$

Substitution of Eq. (3.42) in $\eta_t(\mu) = \lambda^{2t}\alpha^T M_t^{-2}\alpha/(\mu^2 \|\alpha\|^2)$ gives Eq. (3.37).

Let us prove Eqs. (3.38) and (3.39). As $W_t W_t^+ W_t = W_t$ then

$$M_t^{-1} W_t = (I_r - W_t W_t^+) W_t \mu + \sum_{i=0}^{\infty} (-1)^i (W_t^+)^{i+1} W_t \mu^{-i}$$
$$= \sum_{i=0}^{\infty} (-1)^i (W_t^+)^{i+1} W_t \mu^{-i}.$$

Putting in Eq. (3.41)

$$a_k = (-1)^k (W_t^+)^{k+1} W_t \mu^{-k}, \quad b_k = (-1)^k (W_t^+)^{k+1} \mu^{-k}$$

and using the identity $W_t^+ W_t W_t^+ = W_t^+$, we obtain uniformly converging in $t = 1, 2, \ldots, N$ and sufficiently large values of μ power series

$$M_t^{-1} W_t M_t^{-1} = \sum_{i=0}^{\infty} (-1)^i (i+1)(W_t^+)^{i+1} \mu^{-i}$$
$$= W_t^+ - 2(W_t^+)^2/\mu + \sum_{i=0}^{\infty} (-1)^i (i+3)(W_t^+)^{i+3} \mu^{-i-2} \tag{3.43}$$
$$= W_t^+ - 2(W_t^+)^2/\mu + R\sum_{i=0}^{\infty} (-1)^i (i+3)(W_t^+)^{i+1} \mu^{-i-2}.$$

that implies Eqs. (3.38) and (3.39).

Theorem 3.5.

The asymptotic representation

$$\beta_t(\mu) = \eta_t^0 + \gamma_t^0 + \gamma_t^1/\mu + O(\mu^{-2}), \quad \mu \to \infty, \tag{3.44}$$

is valid for $t = 1, 2, \ldots, tr$, where

$$\eta_t^0 = \lambda^{-2t}\alpha^T (I_r - W_t W_t^+)\alpha/\|\alpha\|^2, \quad \gamma_t^0 = R trace(V_t^+)/\|\alpha\|^2,$$

$$\gamma_t^1 = -2\lambda^t R trace(W_t^+ V_t^+)/\|\alpha\|^2, \quad tr = \min_t\{t: V_t > 0, \ t = 1, 2, \ldots, N\},$$

$$V_t = \sum_{k=1}^{t} C_k^T C_k.$$

Proof

Let us present Eq. (2.18) in the following form

$$M_t^{-1} = \lambda^{-t}(I_r - W_t W_t^+)\mu + \Lambda_t W_t^+, \tag{3.45}$$

where

$$\Lambda_t = \sum_{i=0}^{\infty}(-1)^i \lambda^{ti}(W_t^+)^i \mu^{-i}.$$

Using this expression gives

$$M_t^{-1}\sum_{s=1}^{t}\lambda^{t-s}C_s^T\xi_s = \lambda^{-t}(I_r - W_t W_t^+)\sum_{s=1}^{t}\lambda^{t-s}C_s^T\xi_s\mu + \Lambda_t W_t^+\sum_{s=1}^{t}\lambda^{t-s}C_s^T\xi_s.$$

With the help of Lemma 2.2 we establish for $\Xi_t = W_t$, $F_t = \lambda^{-t/2}C_t$ that

$$(I_r - W_t W_t^+)C_s^T = 0, \quad s = 1, 2, \ldots, t.$$

We have

$$W_t = \sum_{s=1}^{t}\lambda^{t-s}C_s^T C_s = \lambda^t \overline{C}_t^T R_t^{-1}\overline{C}_t = \lambda^t \tilde{C}_t^T \tilde{C}_t,$$

$$W_t^+ = \lambda^{-t}(\tilde{C}_t^T\tilde{C}_t)^+\sum_{s=1}^{t}\lambda^{t-s}C_s^T\xi_s = \overline{C}_t^T R_t^{-1/2}\tilde{\xi}_t,$$

where

$$\tilde{C}_t = (C_1^T\lambda^{-1/2}, C_2^T\lambda^{-1}, \ldots, C_t^T\lambda^{-t/2})^T \in R^{t\times r}, \quad \tilde{\xi}_t = (\xi_1, \xi_2, \ldots, \xi_t)^T \in R^t,$$

$$R_t = diag(\lambda, \lambda^2, \ldots, \lambda^t) \in R^{t\times t}, \quad \overline{C}_t = (C_1^T, C_2^T, \ldots, C_t^T)^T.$$

Whence it follows

$$\begin{aligned} M_t^{-1}\sum_{s=1}^{t}\lambda^{t-s}C_s^T\xi_s &= \Lambda_t W_t^+\sum_{s=1}^{t}\lambda^{t-s}C_s^T\xi_s \\ &= \Lambda_t\tilde{C}_t^+ R_t^{-1/2}\tilde{\xi}_t = \Lambda_t\overline{C}_t^+\tilde{\xi}_t, \quad t = 1, 2, \ldots, r. \end{aligned} \tag{3.46}$$

As

$$\overline{C}_t^+(\overline{C}_t^+)^T = (\overline{C}_t^T\overline{C}_t)^+, \quad \Lambda_t = I_r - \lambda^t W_t^+/\mu + O(1/\mu^2)$$

then

$$E\left[M_t^{-1}\sum_{s=1}^{t}\lambda^{t-s}C_s^T\xi_s\sum_{j=1}^{t}\lambda^{t-j}\xi_j C_j M_t^{-1}\right] = R\Lambda_t\overline{C}_t^+(\overline{C}_t^+)^T\Lambda_t = R\Lambda_t V_t^+\Lambda_t$$
$$= RV_t^+ - \lambda^t RW_t^+V_t^+/\mu - \lambda^t RV_t^+W_t^+/\mu + O(1+\mu^2),$$

this implies Eq. (3.44).

Let us establish some important consequences arising from Theorems 3.4 and 3.5.

Keeping in Eq. (3.44) the terms of the zero-order with respect to $1/\mu$, we obtain

$$\beta_t(\mu) = \eta_t^0 + \gamma_t^0 + O(\mu^{-1}), \quad \mu \to \infty, \tag{3.47}$$

for $t = 1, 2, \ldots, N$, where

$$\eta_t^0 = \lambda^{-2t}\alpha^T(I_r - W_t W_t^+)\alpha/||\alpha||^2 \quad \gamma_t^0 = Rtrace(V_t^+)/\|\alpha\|^2.$$

Since the matrix $I_r = W_t W_t^+$ is nonnegative definite and idempotent, then from Eq. (3.47) we find under the assumption that C_t is not a linear combination of $C_1, \ldots, C_{t-1}$, $t = 2, 3, \ldots, tr - 1$

$$\begin{aligned}\beta_t(\mu) &\leq 1 + \|C_t\|^2 trace(V_t^+)/S_t + O(\mu^{-1}) \\ &= 1 + \|C_t\|^2 \sum_{i=1}^{t} \lambda_t^{-1}(i)/S_t + O(\mu^{-1}),\end{aligned}$$

where $\lambda_t(i), i = 1, 2, \ldots, r$ are the eigenvalues of the matrix V_t. This implies that the RLSM overshooting for arbitrarily large values of μ and $t = 1, 2, \ldots, tr$ is bounded in norm. As μ enters singularly in Eq. (3.3), this result is not obvious even though it is considered a finite interval of observation. At the same time, closeness to zero of eigenvalues can lead to arbitrarily large values of the overshoot which will continue over time, significantly exceeding $t = tr$.

Example 3.3.

Consider the problem of the neural network (NN) training from the Example 2.5. Fig. 3.12 shows the dependencies β_t on the count for the realizations of the NN with 10 neurons.

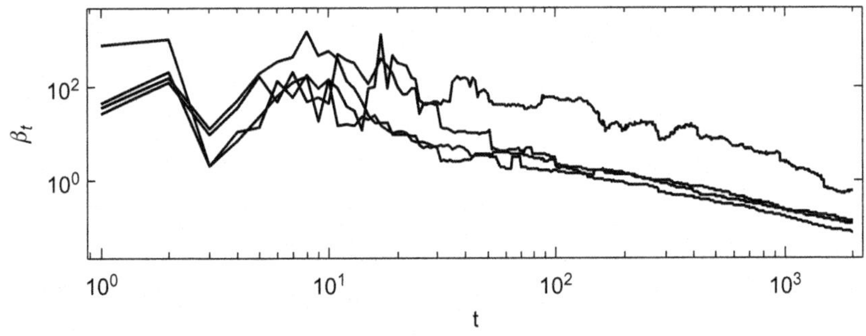

Figure 3.12 Dependency $\beta_t(\mu)$ on t.

From Eq. (3.44) the necessary and sufficient condition for the absence of the overshoot in the zero approximation in the end of the transitional phase follows

$$S_{tr} \geq \| S_{tr} \|^2 trace(V_{tr}^{-1}). \tag{3.48}$$

Example 3.4.
We will illustrate the use of the obtained criterion Eq. (3.48) of the overshoot absence at the end point of the transitional phase for the large signal to noise ratio using the signal model of the Example 3.1.

Consider two cases: $S_t = 43.9$ dB and $S_t = 44$ dB. Since the value of right-hand side of Eq. (3.48) is equal to 43.9 dB, in the first case the overshoot should occur, and in the second case it should be absent for any $\mu > 0$. Fig. 3.13 shows the simulation results supporting the calculations when $\mu = 10^7$, in curves 1 and 2, respectively.

Introduce the notation $\beta_t^{1-\varepsilon}(\mu)$ for the function

$$\beta_t(\mu) = \eta_t^0 + \gamma_t^0 + \gamma_t^1/\mu$$

that takes into account the first order of the smallness with respect to $1/\mu$ for

$$\lambda = 1 - \varepsilon > 0, \quad \varepsilon \geq 0.$$

Let us analyze the sensitivity of $\beta_t(\mu)$ to λ at the moment of the transitional phase end. We have

$$\begin{aligned}\beta_t^{1-\varepsilon}(\mu) &= R[trace(V_{tr}^{-1}) - 2\lambda^{tr} Rtrace(W_{tr}^{-1} V_{tr}^{-1})/\mu]/\|\alpha\|^2 \\ &= R\left[trace(V_{tr}^{-1}) - 2Rtrace\left(\left(\sum_{k=1}^{tr}(1-\varepsilon)^{-k} C_k^T C_k\right)^{-1} V_{tr}^{-1}\right)/\mu\right]\|\alpha\|^2.\end{aligned} \tag{3.49}$$

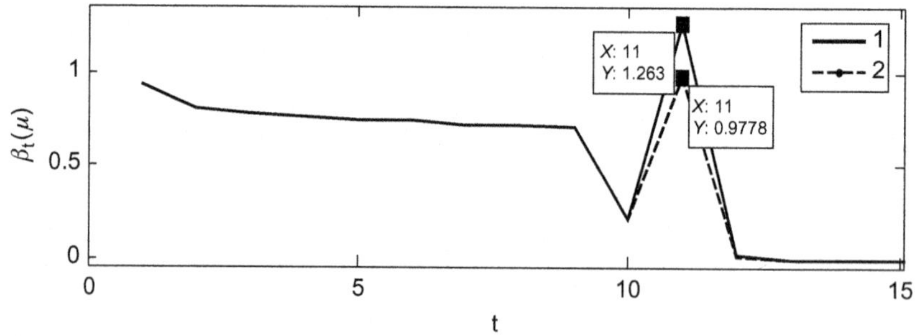

Figure 3.13 Dependency $\beta_t(\mu)$ on t for $\mu = 10^7$, $S_t = 44$ dB (1), $S_t = 43.9$ dB (2).

It follows from this

$$\frac{\partial}{\partial\varepsilon}\beta_t^{1-\varepsilon}(\mu)|_{\varepsilon=0} = 2\|C_{tr}\|^2 trace\left(\sum_{k=1}^{tr} kC_k^T C_k V_{tr}^{-3}\right)/(S_{tr}\mu) > 0. \quad (3.50)$$

Taking into account that

$$V_{tr} \le \sum_{k=1}^{tr} kC_k^T C_k \le trV_{tr}$$

the two-sided estimate

$$\chi \le \frac{\partial}{\partial\varepsilon}\beta_t^{1-\varepsilon}(\mu)|_{\varepsilon=0} \le tr\chi, \quad (3.51)$$

follows from Eq. (3.50), where

$$\chi = 2\|C_{tr}\|^2 trace(V_{tr}^{-2})/(S_{tr}\mu) > 0.$$

3.4 FLUCTUATIONS UNDER DIFFUSE INITIALIZATION

Let us study the fluctuation behavior of the diffuse algorithm which is described by expressions:

$$\alpha_t^{dif} = \alpha_{t-1}^{dif} + K_t^{dif}(y_t - C_t\alpha_{t-1}^{dif}), \quad \alpha_0^{dif} = 0, \quad (3.52)$$

$$K_t^{dif} = W_t^+ C_t^T, \quad (3.53)$$

$$W_t = \lambda W_{t-1} + C_t^T C_t, \quad W_0 = 0_{r\times r}, \quad t = 1, 2, \ldots, N. \quad (3.54)$$

Theorem 3.6.
The representations

$$a_t = E[e_t] = -(I_r - W_t W_t^+)\alpha, \quad (3.55)$$

$$\beta_t = E[e_t^T e_t]/||\alpha||^2 = \eta_t + \gamma_t, \quad t = 1, 2, \ldots, N, \quad (3.56)$$

are valid, where

$$e_t = \alpha_t^{dif} - \alpha, \quad \eta_t = \lambda^{-2t}\alpha^T(I_r - W_t W_t^+)\alpha/||\alpha||^2, \quad \gamma_t = Rtrace(V_t^+)/||\alpha||^2.$$

Proof

Since the bias a_t satisfies the equations system

$$a_t = (I_r - K_t^{dif} C_t)a_{t-1}, a_0 = -\alpha, \quad t = 1, 2, \ldots, N$$

then it follows from Lemma 2.4. the expression for its transition matrix

$$X_{t,0} = I_r - W_t^+ W_t, \quad t = 1, 2, \ldots, N$$

which implies Eq. (3.55).

Iterating the system equations for the estimation error $e_t = \alpha_t - a$

$$e_t = (I_r - K_t^{dif}{}_t C_t)e_{t-1} + K_t \xi_t, e_t = -\alpha, \quad t = 1, 2, \ldots, N,$$

we find

$$e_t = -X_{t,0}\alpha + \sum_{s=1}^{t} X_{t,s} K_t^{dif} \xi_s, \quad t = 1, 2, \ldots, N, \tag{3.57}$$

where $X_{t,s}$ is the transition matrix of the homogeneous system

$$x_t = (I_r - K_t^{dif} C_t)x_{t-1}, \quad t = 1, 2, \ldots, N.$$

Substitution of

$$X_{t,0} = (I_r - W_t W_t^+), \;\; X_{t,s} K_s^{dif} = \lambda^{t-s} W_t^+ C_s^T$$

in Eq. (3.57) gives

$$e_t = -(I_r - W_t W_t^+)\alpha + W_t^+ \sum_{s=1}^{t} \lambda^{t-s} C_s^T \xi_s, \quad t = 1, 2, \ldots, N. \tag{3.58}$$

Since

$$W_t = \sum_{s=1}^{t} \lambda^{t-s} C_s^T C_s = \lambda^t \overline{C}_t^T R_t^{-1} \overline{C}_t = \lambda^t \tilde{C}_t^T \tilde{C}_t,$$

$$W_t^+ = \lambda^{-t} (\tilde{C}_t^T \tilde{C}_t)^+, \sum_{s=1}^{t} \lambda^{t-s} C_s^T \xi_s = \tilde{C}_t^T R_t^{-1/2} \tilde{\xi}_t,$$

where

$$\tilde{C}_t = (C_1^T \lambda^{-1/2}, C_2^T \lambda^{-1}, \ldots, C_t^T \lambda^{-t/2})^T \in R^{t \times r}, \quad \tilde{\xi}_t = (\xi_1, \xi_2, \ldots, \xi_t)^T \in R^t,$$

$$R_t = diag(\lambda, \lambda^2, \ldots \lambda^t) \in R^{t \times t}, \quad \overline{C}_t = (C_2^T, C_1^T, \ldots, C_t^T)^T$$

then

$$W_t^+ \sum_{s=1}^{t} \lambda^{t-s} C_s^T \xi_s = \tilde{C}_t^+ R_t^{-1/2} \tilde{\xi}_t = \overline{C}_t^+ \tilde{\xi}_t,$$

$$E\left[W_t^+ \sum_{s=1}^{t} \lambda^{t-s} C_s^T \xi_s \sum_{j=1}^{t} \lambda^{t-j} \xi_j C_j W_t^+ \right]$$
$$= R\overline{C}_t^+ (\overline{C}_t^+)^T = RV_t^+, \quad t = 1, 2, \ldots, N.$$

This implies Eq. (3.56).

Consider properties of β_t, γ_t, η_t for the RLSM with the diffuse initialization.

Theorem 3.7.

Let C_t be not a linear combination of $C_1, \ldots, C_{t-1}$, $t = 2, 3, \ldots, tr$. Then:

1. γ_t is monotonically increasing function in t if $t = 2, 3, \ldots, tr - 1$ and monotonically decreasing if $t = tr, \quad tr + 1, \ldots, N$.
2. η_t is monotonically decreasing function in t if $t = 2, 3, \ldots, tr - 1$ and it vanishes at $t = tr$.

Proof

1. Let us denote

$$Q_t = \sum_{k=1}^{t} \lambda^{-k} C_k^T C_k = \sum_{k=1}^{t} \tilde{C}_k^T \tilde{C}k,$$

where $\tilde{C}_t = \lambda^{-t/2} C_t$. We use the following recursive definition of Q_t^+ [47]

$$\begin{aligned} Q_t^+ &= Q_{t-1}^+ + \frac{1 + \tilde{C}_t Q_{t-1}^+ \tilde{C}_t^T}{(\tilde{C}_t \Xi_{t-1} \tilde{C}_t^T)^2} (\Xi_{t-1} \tilde{C}_t^T)(\Xi_{t-1} \tilde{C}_t^T)^T - \\ &- \frac{Q_{t-1}^+ \tilde{C}_t^T (\Xi_{t-1} \tilde{C}_t^T)^T + (\Xi_{t-1} \tilde{C}_t^T)(Q_{t-1}^+ \tilde{C}_t^T)^T}{\tilde{C}_t \Xi_{t-1} \tilde{C}_t^T}, \quad t = 2, 3, \ldots, tr \end{aligned} \tag{3.59}$$

with the initial condition

$$Q_1^+ = (\tilde{C}_1^T \tilde{C}_1)^+ = \tilde{C}_1^+ (\tilde{C}_1^T)^+ = \tilde{C}_1^T \tilde{C}_1 / ||\tilde{C}_1||^4,$$

where $\Xi_t = I_r - Q_t Q_t^+$.

From Eq. (3.59) we find taking into account that $Q_{t-1}^+ \Xi_{t-1} = 0$

$$\begin{aligned} trace(Q_t^+) &= trace(Q_{t-1}^+) + \frac{1+\tilde{C}_t Q_{t-1}^+ \tilde{C}_t^T}{\tilde{C}_t \Xi_{t-1} \tilde{C}_t^T} - \frac{2\tilde{C}_t Q_{t-1}^+ \Xi_{t-1} \tilde{C}_t^T}{\tilde{C}_t \Xi_{t-1} \tilde{C}_t^T} \\ &= trace(Q_{t-1}^+) + \frac{1+\tilde{C}_t Q_{t-1}^+ \tilde{C}_t^T}{\tilde{C}_t \Xi_{t-1} \tilde{C}_t^T}, \quad t = 2, 3, \ldots, tr. \end{aligned} \tag{3.60}$$

Since

$$\gamma_t = Rtrace(W_t^+)/\|\alpha\|^2 = \lambda^{-t} Rtrace(Q_t^+)/\|\alpha\|^2$$

then using Eq. (3.60), we obtain

$$\gamma_t = \lambda^{-1}\gamma_{t-1} + \lambda^{-t} R \frac{1+\tilde{C}_t Q_{t-1}^+ \tilde{C}_t^T}{\tilde{C}_t \Xi_{t-1} \tilde{C}_t^T} / \|\alpha\|^2, \quad t = 2, 3, \ldots, tr, \tag{3.61}$$

$$\gamma_1 = \lambda^{-1} R / \|C_1\|^2 / \|\alpha\|^2. \tag{3.62}$$

This implies that γ_t is monotonically increasing function for $t = 2, 3, \ldots, tr - 1$.

For the values $t = tr + 1, \quad tr + 2, \ldots, N$ we use the following recursive definition of Q_t^{-1} [47]

$$Q_t^{-1} = Q_{t-1}^{-1} - \frac{Q_{t-1}^{-1} \tilde{C}_t^T \tilde{C}_t Q_{t-1}^{-1}}{1+\tilde{C}_t Q_{t-1}^{-1} \tilde{C}_t^T}$$

with the initial condition

$$Q_{tr}^{-1} = \left(\sum_{k=1}^{tr} \lambda^{-k} C_k^T C_k \right)^{-1}.$$

The assertion follows from the expressions

$$\gamma_t = \lambda^{-1}\gamma_{t-1} - \lambda^{-t} R \frac{\tilde{C}_t Q_{t-1}^{-2} \tilde{C}_t^T}{1+\tilde{C}_t Q_{t-1}^{-1} \tilde{C}_t^T} / \|\alpha\|^2,$$

$$t = tr + 1, \quad tr + 2, \ldots, N,$$

$$\gamma_{tr} = Rtrace\left(\sum_{k=1}^{t} \lambda^{t-k} C_k^T C_k \right)^{-1} / ||\alpha||^2.$$

2. Let $\lambda_t(i)$ be the eigenvalues of the matrix W_t and $p_t(i)$ are corresponding them eigenvectors, $i = 1, 2, ..., r$, $t = 2, 3, ..., tr$. Assume without loss of generality that

$\lambda_t(i) = 0$ for $i = 1, 2, \ldots, r - t$ and $\lambda_t(i) > 0$, $i = r - t + 1, r - t + 2, \ldots, r$.

We have

$$W_t = \sum_{i=1}^{r} \lambda_t(i) p_t(i) p_t^T(i), \quad W_t^+ = \sum_{i=1}^{r} \lambda_t^+(i) p_t(i) p_t^T(i).$$

Since $p_t(i)$, $i = 1, 2, ..., r$ are the orthogonal vectors of unit length for each $t = 1, 2, \ldots, tr - 1$, we have

$$W_t W_t^+ = \sum_{i=1}^{r} \sum_{j=1}^{r} \lambda_t(i) \lambda_t^+(j) p_t(i) p_t^T(j)$$

$$= \sum_{i=1}^{r} \lambda_t(i) \lambda_t^+(j) p_t(i) p_t^T(i) = \sum_{=r-t+1}^{r} p_t(i) p_i^T(i)$$

$$I_r - W_t W_t^+ = \sum_{i=1}^{r-t} p_t(i) p_t^T(i).$$

From this it follows that

$$\eta_t = \lambda^{2t} \sum\nolimits_{i=1}^{r-t} (\alpha^T p_t(i))^2 / ||\alpha||^2$$

and at the same time the assertion of the theorem.

Theorem 3.8.
Let $\lambda = 1$, C_t be not a linear combination of $C_1, \ldots, C_{t-1}$, $t = 2, 3, ..., tr$ and the signal to noise ratio S_t satisfies the inequality

$$S_t < 1 + C_t W_{t-1}^+ C_t^T, \quad t = 2, 3, \ldots, tr - 1. \tag{3.63}$$

Then β_t is a monotonically increasing function in t for $t = 2, 3, \ldots, tr$.

Proof
Using the relations

$$\Xi_t = \Xi_{t-1} - \frac{\Xi_{t-1} C_t^T C_t \Xi_{t-1}}{C_t \Xi_{t-1} C_t^T}, \quad \Xi_1 = I_r, \quad t = 2, 3, \ldots, tr$$

to determine the idempotent matrix $\Xi_t = I_r - W_t W_t^+$ [47], we obtain the following recursive form presentation of $\eta_t = \alpha^T (I_r - W_t W_t^+) \alpha / \|\alpha\|^2$:

$$\eta_t = \eta_{t-1} - \frac{(C_t\Xi_{t-1}\alpha)^2}{C_t\Xi_{t-1}C_t^T} / \|\alpha\|^2, \quad \eta_0 = 1, \quad t = 2, 3, \ldots, tr. \tag{3.64}$$

Since $\beta_t = \eta_t + \gamma_t$ we find from Eqs. (3.61), (3.62), and (3.64)

$$\beta_t = \beta_{t-1} + \frac{R(1 + C_t W_{t-1}^{+} C_t^T) - (C_t\Xi_{t-1}\alpha)^2}{\tilde{C}_t\Xi_{t-1}\tilde{C}_t^T} / \|\alpha\|^2, \quad t = 2, 3, \ldots, tr, \tag{3.65}$$

$$\beta_1 = 1 + R/\|C_1\|/\|\alpha\|^2. \tag{3.66}$$

This implies that β_t will monotonically increasing function for $t = 2, 3, \ldots, tr$ if the following condition is fulfilled

$$(C_t\Xi_{t-1}\alpha)^2 \le R(1 + C_t W_{t-1}^{+} C_t^T)_t^T, \quad t = 2, 3, \ldots, tr. \tag{3.67}$$

We show that

$$(C_t\Xi_{t-1}\alpha)^2 \le \|C_t\|^2 \|\alpha\|^2. \tag{3.68}$$

Let $B, D \in R^{n \times n}$ be arbitrary, symmetric matrice eigenvalues which satisfy the condition

$$\lambda_i(B) \ge \lambda_i(D), \quad i = 1, 2, \ldots, n.$$

Then there is an orthogonal matrix $T \in R^{n \times n}$ such that [67]

$$T^T B T \ge D.$$

Let us denote

$$B = C_t^T C_t, \quad D = \Xi_{t-1} C_t^T C_t \Xi_{t-1}.$$

Since the non-zero eigenvalues of B, D are determined by expressions

$$\lambda(B) = C_t C_t^T, \quad \lambda(D) = C_t \Xi_{t-1}^2 C_t^T = C_t \Xi_{t-1} C_t^T,$$

we have $\lambda(B) \ge \lambda(D)$ and $T^T B T \ge D$. Thus

$$(C_t\Xi_{t-1}\alpha)^2 = \alpha^T \Xi_{t-1} C_t^T C_t \Xi_{t-1} \alpha \le \alpha^T T^T C_t^T C_t T \alpha \le$$

$$\le C_t T T^T C_t^T \alpha^T \alpha = \|C_t\|^2 \|\alpha\|^2$$

Using Eq. (3.66) together with Eq. (3.68) gives sufficient condition for β_t to be a monotonically increasing function in t

$$\| C_t \|^2 \| \alpha \|^2 \leq R(1 + C_t W_{t-1}^+ C_t^T), \quad t = 1, 2, \ldots, tr.$$

But since $S_t = \| C_t \|^2 \| \alpha \|^2 / R$ then this implies the theorem assertion.

3.5 FLUCTUATIONS WITH RANDOM INPUTS

Consider the fluctuations of the RLSM with random inputs C_t. Restrict ourselves to the case of the diffuse initialization. We have

$$e_t = \alpha_t - \alpha = -H_{t,0}\alpha + \sum_{s=1}^{t} H_{t,s} K_s^{dif} \xi_s, \quad t = 1, 2, \ldots, N,$$

where $H_{t,s}$ is the transition matrix of the system Eq. (3.10).

Since $H_{t,0} = I_r - W_t^+ W_t$ then assuming that C_t does not depend on ξ_t for any $t = 1,2,\ldots,N$, we get when $t \geq tr$

$$E[e_t] = E\{E\left[-H_{t,0}\alpha + \sum_{s=1}^{t} H_{t,s} K_s^{dif} \xi_s\right] | C_s, s = 1, 2, \ldots, t\} = 0,$$

where $E(\zeta | \upsilon)$ is conditional expectation of a random variable ζ for a given other random variable υ. Thus, as in the case of deterministic input signals, the bias in the diffuse algorithm is absent if $t \geq tr$.

For conditional matrix of second moments of the estimation error with the help of Lemma 3.5. and the relations established in the proof of Theorem 3.6, we find

$$E[e_t e_t^T | C_s, s = 1, 2, \ldots, tr] = \lambda^{2t}(I_r - W_t W_t^+)\alpha^T \alpha (I_r - W_t W_r^+) + R V_t^+,$$

where $V_t = \sum_{k=1}^{t} C_k^T C_k$. Thus

$$E[e_t^T e_t | C_s, s = 1, 2, \ldots, tr] = \lambda^{2t} \alpha^T (I_r - W_t W_t^+)\alpha + R trace(V_t^+). \quad (3.69)$$

Assume that the input signals have the property of ergodicity, i.e., for sufficiently large tr the approximation

$$E\left(\sum_{k=1}^{t} C_k^T C_k\right) \approx \Sigma = diag(\sigma^2(1), \sigma^2(2), \ldots, \sigma^2(r)), \quad (3.70)$$

is possible, where $\sigma^2(i) = E[C_i C_i^T]$, $i = 1,2,\ldots,r$. Then from Eq. (3.69) it follows that

$$\begin{aligned}\beta_{tr} &= E\{E[e_{tr}^T e_{tr}|C_s, s=1,2,\ldots,tr]\}/\|\alpha\|^2 = R\textstyle\sum_{i=1}^{r} 1/\sigma^2(i)/\|\alpha\|^2 \\ &= \sum_{i=1}^{r}\sigma^2(i)\sum_{i=1}^{r}1/\sigma^2(i)/S,\end{aligned} \tag{3.17}$$

where $S = \sum_{i=1}^{r}\sigma^2(i)/\|\alpha\|^2/R$ is the ratio of signal to noise.

From whence we find the condition for the absence of overshoot at the moment of the end of the transitional stage

$$S \geq \sum_{i=1}^{r}\sigma^2(i)\sum_{i=1}^{r}1/\sigma^2(i). \tag{3.72}$$

For the case $\sigma^2(i) = \sigma^2,\ i = 1,2,\ldots,r$ it has particularly simple form $S \geq r^2$.

We obtain now an analogue of Eq. (3.72) under the assumption that $C_1^T, C_2^T, \ldots, C_N^T$ is a sequence of independent, identically distributed random vectors with multivariate normal distribution $N(0,\Sigma),\ \Sigma > 0$, $t = 1,2,\ldots,N$. It is known that in this case

$$\left(\sum_{k=1}^{t} C_k^T C_k\right)^{-1}$$

has the inverse distribution Wishart [67] when $t \geq r+2$ with probability 1 and

$$E = \left[\left(\sum_{k=1}^{t} C_k^T C_k\right)^{-1}\right] = \Sigma^{-1}$$

[68]. Therefore, the use of Eq. (3.69) for $t = r+2$ gives

$$\begin{aligned}\beta_{r+2} &= E\{E[e_{r+2}^T e_{r+2}|C_s, s=1,2,\ldots,r+2]\}/\|\alpha\|^2 = \\ &= Rttrace(\Sigma^{-1})/||\alpha^2|| = Rtrace(\Sigma)trace(\Sigma^{-1})/S,\end{aligned} \tag{3.73}$$

where $S = \|\alpha\|^2\ trace(\Xi)/R$ is the ratio of signal to noise.

From whence we find the condition for overshoot absence at the end moment of the transitional stage

$$S \geq trace(\Sigma)trace(\Sigma^{-1}). \tag{3.74}$$

It is seen that under the assumption of correlations absence between inputs this condition coincides with Eq. (3.72).

Example 3.5.

We will illustrate the condition of the overshoot absence at the moment $t = r + 2$ when with probability 1 the matrix W_t is nonsingular. Consider the observation model Eq. (3.1) with random outcomes. We assume that

$$E[C_t(i)] = 0, \quad E[C_t^2(i)] = \sigma^2 = 1, \quad i = 1, 2, \ldots, r, \quad t = 1, 2, \ldots, r + 2,$$
$$\alpha_i = 0.1i, \quad i = 1, 2, \ldots, r.$$

The evaluation results using a statistical modeling of 500 realizations are shown in Table 3.1.

Table 3.1 Estimates of β_{tr+2} obtained by means of the statistical simulation

r	5	10	20	30	40	50	5	10	20	30	40	50
S	122	127	434	982	1830	2552	18	78.6	340	709	1310	1411
S, dB	20.9	21	26.4	30	33.6	34	13.6	19	25	28.5	31.5	13.6
β_{tr+2}	0.19	0.7	0.86	0.73	0.73	0.76	1.58	1.17	1.26	1.26	1.33	1.7
$R^{1/2}$	0.15	0.55	1.15	1.7	3.2	3.9	0.39	0.7	1.3	3.0	3.6	3.9

In the left half of the table the noise intensity was chosen so to ensure fulfillment of the condition $S \geq r^2$ under which the overshoot is absent and in the right, so that this condition is not fulfilled. It can be seen that the simulation matches the obtained theoretical results.

CHAPTER 4

Diffuse Neural and Neuro-Fuzzy Networks Training Algorithms

Contents

4.1 PROBLEM STATEMENT

Consider an observation model of the form

$$y_t = \Phi(z_t, \beta)\alpha + \xi_t, \quad t = 1, 2, \ldots, N, \tag{4.1}$$

where $z_t \in R^n$ is a vector of inputs, $y_t \in R^m$ is a vector of outputs, $\alpha \in R^r$ and $\beta \in R^l$, are vectors of unknown parameters, $\Phi(z_t, \beta)$ is an $m \times r$ matrix of given nonlinear functions, N is a training set size, and $\xi_t \in R^m$ is a random process which has uncorrelated values, zero expectation, and a covariance matrix R_t.

We suppose that the following conditions are satisfied for the parameters β and α appearing in description Eq. (4.1).

A1: For the vector β, its probable value $\overline{\beta}$ is known and possible deviations from $\overline{\beta}$ are characterized by the function

$$\Gamma(\beta) = (\beta - \overline{\beta})^T \overline{P}_\beta^{-1} (\beta - \overline{\beta}), \tag{4.2}$$

Diffuse Algorithms for Neural and Neuro-Fuzzy Networks.
DOI: http://dx.doi.org/10.1016/B978-0-12-812609-7.00004-4

where $\overline{P}_\beta > 0$ is a given positive definite matrix.
A2: A priori information on elements of the vector α is absent and they are interpreted as random variables with

$$E(\alpha) = 0, \quad E(\alpha\alpha^T) = \mu\overline{P}_\alpha, \tag{4.3}$$

where $\mu > 0$ is a large parameter selected in the course of simulation, $\overline{P}_\alpha > 0$ is a positive definite $r \times r$ matrix independent of μ.
A3: The vector α is not correlated with β and ξ_t, $t = 1, 2, \ldots, N$.

It is assumed that $\overline{\beta}$ and $\overline{P}_\beta$ can be obtained from a training set, the distribution of a generating set or some linguistic information.

Let the training set be specified $\{z_t, y_t\}$, $t = 1, 2, \ldots, N$ and let the quality criterion at a moment t be defined by the expression

$$\begin{aligned} J_t(x) = \sum_{k=1}^{t} \lambda^{t-k}(y_k - \Phi(z_k, \beta)\alpha)^T(y_k - \Phi(z_k, \beta)\alpha) \\ + \lambda^t\Gamma(\beta) + \lambda^t\alpha\overline{P}_\alpha^{-1}\alpha/\mu, \quad t = 1, 2, \ldots, N. \end{aligned} \tag{4.4}$$

A vector of parameters $x = (\beta^T, \alpha^T)^T \in R^{l+r}$ is found under the condition that $J_t(x)$ is minimal and the result must be updated after obtaining a new observation. We use the GN method with linearization in the neighborhood of the last estimate to solve this problem and study its behavior for a large μ (soft initialization) and as $\mu \to \infty$ (diffuse initialization).

Consider an alternative approach to the training problem in the absence of a priori information about α. In this approach it is assumed that the unknown parameters in Eq. (4.1) satisfy the following conditions:
B1: For the vector β, its probable value $\overline{\beta}$ is known and possible deviations from $\overline{\beta}$ are characterized by the function Eq. (4.2).
B2: A priori information on elements of the vector α is absent and they can be either unknown constants or random quantities whose statistical characteristics are unknown.
Let the training set be specified $\{z_t, y_t\}$, $t = 1, 2, \ldots, N$ and let the quality criterion at a moment t be defined by the expressions

$$J_t(x) = \sum_{k=1}^{t} \lambda^{t-k}(y_k - \Phi(z_k, \beta)\alpha)^T(y_k - \Phi(z_k, \beta)\alpha) + \lambda^t\Gamma(\beta). \tag{4.5}$$

A vector of parameters $x = (\beta^T, \alpha^T)^T \in R^{l+r}$ is found under the condition that $J_t(x)$ is minimal and the result must be updated after obtaining a new observation. As above, we use the GN method with linearization in the neighborhood of the last estimate to solve this problem.

4.2 TRAINING WITH THE USE OF SOFT AND DIFFUSE INITIALIZATIONS

Let us assume that the unknown parameters in Eq. (4.1) satisfy conditions A1, A2, and A3 and the quality criteria at a moment t is defined by Eq. (4.4).

Lemma 4.1.

The solution of the minimization problem (4.4) by the GN method with the soft initialization can be found recursively as follows

$$x_t = x_{t-1} + K_t(y_t - h_t(x_{t-1})), \quad x_0 = (\overline{\beta}^T, 0_{1\times r})^T, \quad t = 1, 2, \ldots, N, \quad (4.6)$$

where $x_t = (\beta_t^T, \alpha_t^T)^T$, $h_t(x_{t-1}) = \Phi(z_t, \beta_{t-1})\alpha_{t-1}$,

$$K_t = ((K_t^\beta)^T, (K_t^\alpha)^T)^T = P_t C_t^T, \quad (4.7)$$

$$K_t^\beta = (S_t + V_t L_t V_t^T)(C_t^\beta)^T + V_t L_t (C_t^\alpha)^T, \quad (4.8)$$

$$K_t^\alpha = L_t V_t^T (C_t^\beta)^T + L_t (C_t^\alpha)^T, \quad (4.9)$$

$$P_t = \tilde{S}_t + \tilde{V}_t L_t \tilde{V}_t^T, \quad (4.10)$$

$$\tilde{S}_t = \begin{pmatrix} S_t & 0_{l\times r} \\ 0_{r\times l} & 0_{r\times r} \end{pmatrix}, \tilde{V}_t = \begin{pmatrix} V_t \\ I_r \end{pmatrix}, \quad (4.11)$$

$$S_t = S_{t-1}/\lambda - S_{t-1}(C_t^\beta)^T(\lambda I_m + C_t^\beta S_{t-1}(C_t^\beta)^T)^{-1} C_t^\beta S_{t-1}/\lambda, \quad S_0 = \overline{P}_\beta, \quad (4.12)$$

$$\begin{aligned} V_t = {} & (I_l - S_{t-1}(C_t^\beta)^T(\lambda I_m + C_t^\beta S_{t-1}(C_t^\beta)^T)^{-1} C_t^\beta) V_{t-1} \\ & - S_{t-1}(C_t^\beta)^T(\lambda I_m + C_t^\beta S_{t-1}(C_t^\beta)^T)^{-1} C_t^\alpha, \quad V_0 = 0_{l\times r}, \end{aligned} \quad (4.13)$$

$$\begin{aligned} L_t^{-1} = {} & \lambda L_{t-1}^{-1} + \lambda(C_t^v V_t + C_t^\alpha)^T(\lambda I_m + C_t^\beta S_{t-1}(C_t^\beta)^T)^{-1} \\ & \times (C_t^v V_t + C_t^\alpha), \quad L_0^{-1} = \overline{P}_\alpha^{-1}/\mu, \end{aligned} \quad (4.14)$$

$$C_t = (C_t^{\beta}, C_t^{\alpha}),$$
$$C_t^{\beta} = C_t^{\beta}(x_{t-1}) = \partial[\Phi(z_t, \beta_{t-1})\alpha_{t-1}]/\partial\beta_{t-1}, \quad C_t^{\alpha} = C_t^{\alpha}(x_{t-1}) = \Phi(z_t, \beta_{t-1}). \tag{4.15}$$

Proof

Consider the auxiliary linear observation model

$$y_k = C_k^{\beta}\beta + C_k^{\alpha}\alpha + \xi_t, \quad t = 1, 2, \ldots, N, \tag{4.16}$$

and the optimization problem

$$(\alpha^*, \beta^*) = \text{argmin} J_t(\alpha, \beta), \ \ \alpha \in R^r, \ \ \beta \in R^l, \tag{4.17}$$

where

$$J_t(x) = \sum_{k=1}^{t} \lambda^{t-k}(y_k - C_k^{\beta}\beta - C_k^{\alpha}\alpha)^T(y_k - C_k^{\beta}\beta - C_k^{\alpha}\alpha) + \lambda^t\Gamma(\beta) + \lambda^t\alpha\overline{P}^{-1}\alpha/\mu, \quad t = 1, 2, \ldots, N, \tag{4.18}$$

$$y_t \in R^m, \ \ C_t^{\beta} \in R^{m \times l}, \ \ C_t^{\alpha} \in R^{m \times r}, \ \ \xi_t \in R^m.$$

Using the notation x_t for (α^*, β^*), we find

$$x_t = x_{t-1} + K_t(y_t - C_t x_{t-1}), \ \ x_0 = (\overline{\beta}^T, 0_{1 \times r})^T, \quad t = 1, 2, \ldots, N, \tag{4.19}$$

where $C_t = (C_t^{\beta}, C_t^{\alpha})$,

$$K_t = P_t \, C_t^T, \tag{4.20}$$

$$P_t = (P_{t-1} - P_{t-1}C_t^T(\lambda I_m + C_t P_{t-1} C_t^T)^{-1} C_t P_{t-1})/\lambda, P_0 = block \ diag(\overline{P}_{\beta}, \overline{P}_{\alpha}\mu). \tag{4.21}$$

Let us show that the expressions (4.6)–(4.15) define the solution of the problems (4.17) and (4.18) for arbitrary C_t^{β}, C_t^{α}, and $\mu > 0$

Consider two solutions $\tilde{P}_t$ and $\tilde{S}_t$ of the matrix equation

$$P_t = (P_{t-1} - P_{t-1}C_t^T(\lambda I_m + C_t P_{t-1} C_t^T)^{-1} C_t P_{t-1})/\lambda, \tag{4.22}$$

with initial conditions

$$\tilde{P}_0 = block \ diag(\overline{P}_{\beta}, \overline{P}_{\alpha}\mu), \ \ \tilde{S}_0 = block \ diag(\overline{P}_{\beta}, 0_{r \times r}).$$

Since

$$\tilde{S}_0 C_1^T = \begin{pmatrix} \overline{P}_\beta (C_1^\beta)^T \\ 0 \end{pmatrix}$$

then the first block representation in Eq. (4.11) is valid.

Denote $\overline{A} = I_{r+l}/\lambda^{1/2}$, $\overline{C}_t = C_t/\lambda^{1/2}$. Then Eq. (4.22) will take the form

$$P_t = \overline{A}P_{t-1}\overline{A} - \overline{A}P_{t-1}\overline{C}_t^T(I_m + \overline{C}_t P_{t-1}\overline{C}_t^T)^{-1}\overline{C}_t P_{t-1}\overline{A}. \tag{4.23}$$

The difference

$$Q_t = \tilde{P}_t - \tilde{S}_t, \quad t = 1, 2, \ldots, N$$

satisfies the matrix equation [69]

$$Q_t = \tilde{A}_t Q_{t-1}\tilde{A}_t^T - \tilde{A}_t Q_{t-1}\overline{C}_t^T(I_m + \overline{C}_t\tilde{P}_{t-1}\overline{C}_t^T)^{-1}\overline{C}_t Q_{t-1}\tilde{A}_t^T, \tag{4.24}$$

$$Q_0 = \tilde{P}_0 - \tilde{S}_0 = \tilde{P}_0 = block\ diag(0_{l\times l}, \overline{P}_\alpha \mu), \quad t = 1, 2, \ldots, N,$$

where $\tilde{A}_t = \overline{A} - \overline{A}\tilde{S}_{t-1}\overline{C}_t^T(I_m + \overline{C}_t\tilde{S}_{t-1}\overline{C}_t^T)^{-1}\overline{C}_t$.

Returning to the original notation, we obtain

$$Q_t = (A_t Q_{t-1}A_t^T - A_t Q_{t-1}C_t^T(\lambda I_m + C_t\tilde{P}_{t-1}C_t^T)^{-1}C_t Q_{t-1}A_t^T)/\lambda, \tag{4.25}$$

$$Q_0 = \tilde{P}_0 - \tilde{S}_0 = \tilde{P}_0 = block\ diag(0_{l\times l}, \overline{P}_\alpha \mu), \quad t = 1, 2, \ldots, N,$$

where $A_t = I_{r+l} - \tilde{S}_{t-1}C_t^T(\lambda I_m + C_t\tilde{S}_{t-1}C_t^T)^{-1}C_t$.

Let us show that

$$Q_t = \tilde{V}_t L_t \tilde{V}_t^T. \tag{4.26}$$

Substituting this expression in Eq. (4.25), we get

$$\begin{aligned} \tilde{V}_t L_t \tilde{V}_t^T = \Big(& A_t\tilde{V}_{t-1}L_{t-1}\tilde{V}_{t-1}^T A_t^T - A_t\tilde{V}_{t-1}L_{t-1}\tilde{V}_{t-1}^T C_t^T \\ & \times (\lambda I_m + C_t\tilde{P}_{t-1}C_t^T)^{-1}C_t\tilde{V}_{t-1}L_{t-1}\tilde{V}_{t-1}^T A_t^T\Big)/\lambda. \end{aligned} \tag{4.27}$$

This relation will be carried out if the L_t and $\tilde{V}_t$ satisfy the matrix equations

$$\tilde{V}_t = A_t\tilde{V}_{t-1}, \quad \tilde{V}_0 = (0_{l\times r}, I_r)^T, \tag{4.28}$$

$$L_t = (L_{t-1} - L_{t-1}\tilde{V}_{t-1}^T C_t^T(\lambda I_m + C_t\tilde{P}_{t-1}C_t^T)^{-1}C_t\tilde{V}_{t-1}L_{t-1})/\lambda, \quad L_0 = \mu I_r. \tag{4.29}$$

Since

$$\tilde{S}_{t-1}C_t^T = \begin{pmatrix} S_{t-1}(C_t^\beta)^T \\ 0_{r\times l} \end{pmatrix}, \quad C_t\tilde{S}C_t^T = C_t^\beta S_{t-1}(C_t^\beta)^T,$$

$$A_t = I_{r+l} - \begin{pmatrix} S_{t-1}(C_t^\beta)^T N_t^{-1} C_t^\beta & S_{t-1}(C_t^\beta)^T N_t^{-1} C_t^\alpha \\ 0_{r\times l} & 0_{r\times r} \end{pmatrix},$$

where $N_t = \lambda I_m + C_t^\beta S_{t-1}(C_t^\beta)^T$ then the second expression in Eq. (4.11) follows from this and the Eq. (4.28).

Let us transform Eq. (4.29) using the matrix identity Eq. (2.7) for

$$B = L_{t-1}, \quad C = \tilde{V}_{t-1}^T C_t^T, \quad D = \lambda I_m + C_t^\beta S_{t-1}(C_t^\beta)^T.$$

Taking into account that

$$\tilde{P}_t = Q_t + \tilde{S}_t = \tilde{V}_t L_t \tilde{V}_t^T + \tilde{S}_t,$$

we find

$$\begin{aligned} L_t^{-1} &= (L_{t-1} - L_{t-1}\tilde{V}_{t-1}^T C_t^T(\lambda I_m + C_t\tilde{P}_{t-1}C_t^T)^{-1}C_t\tilde{V}_{t-1}L_{t-1})^{-1}\lambda \\ &= (L_{t-1} - L_{t-1}\tilde{V}_{t-1}^T C_t^T(\lambda I_m + C_t^\beta S_{t-1}(C_t^\beta)^T + C_t\tilde{V}_{t-1}L_{t-1}\tilde{V}_{t-1}^T C_t^T)^{-1}C_t\tilde{V}_{t-1}L_{t-1})^{-1}\lambda \\ &= \lambda L_{t-1}^{-1} + \lambda\tilde{V}_{t-1}^T C_t^T(\lambda I_m + C_t\tilde{S}_{t-1}C_t^T)^{-1}C_t\tilde{V}_{t-1}. \end{aligned}$$

But as $C_t\tilde{V}_{t-1} = C_t^\beta V_{t-1} + C_t^\alpha$ then Eq. (4.14) follows from this.

Let us return to the solution of the original nonlinear minimization problem with the criterion Eq. (4.4). Residual linearization $e_t = y_t - \Phi(z_t, \beta)\alpha$ around the point x_{t-1} gives

$$\begin{aligned} e_t = y_t &- \Phi(z_t, \beta_{t-1})\alpha_{t-1} - \Phi(z_t, \beta_{t-1})(\alpha - \alpha_{t-1}) \\ &- \partial[\Phi(z_t, \beta_{t-1})\alpha_{t-1}]/\partial\beta_{t-1}(\beta - \beta_{t-1}) = \tilde{y}_t - C_t x, \end{aligned}$$

where $x = (\alpha^T, \beta^T)^T$,

$$\begin{aligned} \tilde{y}_t &= y_t + \partial[\Phi(z_t, \beta_{t-1})\alpha_{t-1}]/\partial\beta_{t-1}\beta_{t-1} \\ &= y_t - \Phi(z_t, \beta_{t-1})\alpha_{t-1} + C_t x_{t-1}. \end{aligned}$$

Substitution of this expression into criterion Eq. (4.4) leads to the linear optimization problem solution which as shown is defined by the expressions (4.19) and (4.7)–(4.15). This implies the assertion of the lemma.

Lemma 4.2.

The matrices P_t and K_t in Eqs. (4.20) and (4.21) can be expanded in the power series

$$\begin{aligned} P_t &= \lambda^{-t}\tilde{V}_t\overline{P}_\alpha(I_r - W_tW_t^+)\tilde{V}_t^T\mu + \tilde{S}_t + \tilde{V}_tW_t^+\tilde{V}_t^T \\ &+ \sum_{i=1}^{q}(-1)^i\lambda^{ti}\tilde{V}_t(W_t^+\overline{P}_\alpha^{-1})^{i+1}\overline{P}_\alpha\tilde{V}_t^T\mu^{-i} + O(\mu^{-q-1}), \end{aligned} \tag{4.30}$$

$$\begin{aligned} K_t &= [\tilde{S}_t + \tilde{V}_tW_t^+\overline{P}_\alpha^{-1}(I_r + \overline{P}_\alpha^{-1}W_t^+/\mu)^{-1}\tilde{V}_t^T]\overline{P}_\alpha C_t^TR_t^{-1} \\ &= K_t^{dif} + \sum_{i=1}^{q}(-1)^i\lambda^{ti}\tilde{V}_t(W_t^+\overline{P}_\alpha^{-1})^{i+1}\overline{P}_\alpha\tilde{V}_t^TC_t^TR_t^{-1}\mu^{-i} + O(\mu^{-q-1}) \end{aligned} \tag{4.31}$$

which converge uniformly in $t \in T = \{1, 2, \ldots, N\}$ for bounded T and sufficiently large values of μ, where

$$\begin{aligned} W_t &= \lambda W_{t-1} + \lambda(C_t^\beta V_{t-1} + C_t^\alpha)^T(\lambda I_m + C_t^\beta S_{t-1}(C_t^\beta)^T)^{-1}(C_t^\beta V_{t-1} + C_t^\alpha), \\ W_0 &= 0_{r\times r}, \end{aligned} \tag{4.32}$$

$$K_t^{dif} = (\tilde{S}_t + \tilde{V}_tW_t^+\tilde{V}_t^T)C_t^T. \tag{4.33}$$

Proof

It follows from Eq. (4.14) that

$$\begin{aligned} L_t^{-1} &= \sum_{k=1}^{t}\lambda^{t-k+1}(C_t^vV_t + C_t^\alpha)^T(\lambda I_m + C_t^\beta S_{t-1}(C_t^\beta)^T)^{-1}(C_t^vV_t + C_t^\alpha) + \lambda^t\overline{P}_\alpha^{-1}/\mu \\ &= \lambda^t\left(\sum_{k=1}^{t}\lambda^{-k+1}(C_t\tilde{V}_t)^T(\lambda I_m + C_t^\beta S_{t-1}(C_t^\beta)^T)^{-1}C_t\tilde{V}_t + \overline{P}_\alpha^{-1}/\mu\right). \end{aligned}$$

By means of Lemma 2.1 with

$\Omega_t = L_t^{-1}$, $\Omega_0 = \overline{P}_\alpha^{-1}$, $F_t = \lambda^{-(t+1)/2}(\lambda I_m + C_t^\beta S_{t-1}(C_t^\beta)^T)^{-1/2}C_t\tilde{V}_t$

we obtain the power series

$$L_t = \lambda^{-t}\overline{P}_\alpha(I_r - W_tW_t^+)\mu + W_t^+ + \sum_{i=1}^{q}(-1)^i\overline{P}_\alpha^{1/.2}\lambda^{ti}(\overline{P}_\alpha^{-1/.2}W_t^+\overline{P}_\alpha^{-1/.2})^{i+1}\overline{P}_\alpha^{1/2}\mu^{-i} + O(\mu^{-q-1})$$

which converges uniformly in $t\in T$ for bounded T and $\mu > 1/\lambda_{\min}$,

$$\lambda_{\min} = \min\{\lambda_t(i) > 0, t\in T, i = 1, 2, \ldots, r\},\quad \lambda_t(i),\quad t\in T,\quad i = 1, 2, \ldots, r,$$

where $\lambda_t(i)$, $t\in T$, $i = 1, 2, \ldots, r$ are eigenvalues of the matrix $\overline{P}_\alpha^{1/2}W_t\overline{P}_\alpha^{1/2}$. Substituting this representation in $P_t = \tilde{S}_t + \tilde{V}_tL_t\tilde{V}_t^T$ gives Eq. (4.30).

Let us show now that

$$\tilde{V}_t\overline{P}_\alpha(I_r - W_tW_t^+)\tilde{V}_t^T C_t^T = 0. \tag{4.34}$$

Taking in account that

$$W_t = \sum_{k=1}^{t}\lambda^{t-k+1}(C_k^\beta V_{k-1} + C_k^\alpha)^T(\lambda I_m + C_k^\beta S_{k-1}(C_k^\beta)^T)^{-1}(C_k^\beta V_{k-1} + C_k^\alpha)$$
$$= \sum_{k=1}^{t}\lambda^{t-k+1}(C_k\tilde{V}_{k-1})^T(\lambda I_m + C_k^\beta S_{k-1}(C_k^\beta)^T)^{-1}C_k\tilde{V}_{k-1}$$

and using Lemma 2.2 with

$$F_t = \lambda^{-(t+1)/2}(\lambda I_m + C_t^\beta S_{t-1}(C_t^\beta)^T)^{-1/2}C_t\tilde{V}_{t-1}$$

yields

$$\overline{P}_\alpha(I_r - W_tW_t^+)(C_t\tilde{V}_{t-1})^T(\lambda I_m + C_t^\beta S_{t-1}(C_t^\beta)^T)^{-1/2} = 0.$$

This implies

$$\tilde{V}_t\overline{P}_\alpha(I_r - W_tW_t^+)\tilde{V}_t^T A_t^T C_t^T = \tilde{V}_t\overline{P}_\alpha(I_r - W_tW_t^+)\tilde{V}_{t-1}^T C_t^T$$
$$- \tilde{V}_t\overline{P}_\alpha(I_r - W_tW_t^+)\tilde{V}_{t-1}^T C_t^T(\lambda I_m + C_t\tilde{S}_{t-1}C_t^T)^{-1}C_t\tilde{S}_{t-1}C_t^T = 0.$$

Theorem 4.1.
The solution of the minimization problem (4.4) by the GN method with the diffuse initialization can be found recursively as follows

$$x_t^{dif} = x_{t-1}^{dif} + K_t^{dif}(y_t - h_t(x_{t-1}^{dif})),\quad x_0^{dif} = (\overline{\beta}^T, 0_{1\times r})^T,\quad t = 1, 2, \ldots, N, \tag{4.35}$$

where $x_t^{dif} = ((\beta_t^{dif})^T, (\beta_t^{dif})^T)^T$, $K_t^{dif} = ((K_t^{\beta})^T, (K_t^{\alpha})^T)^T$, $h_t(x_{t-1}) = \Phi(z_t, \beta_{t-1})\alpha_{t-1}$,

$$K_t^{\beta} = (S_t + V_t W_t^+ V_t^T)(C_t^{\beta})^T + V_t W_t^+ (C_t^{\alpha})^T, \tag{4.36}$$

$$K_t^{\alpha} = W_t^+ V_t^T (C_t^{\beta})^T + W_t^+ (C_t^{\alpha})^T, \tag{4.37}$$

$$S_t = S_{t-1}/\lambda - S_{t-1}(C_t^{\beta})^T(\lambda I_m + C_t^{\beta} S_{t-1}(C_t^{\beta})^T)^{-1} C_t^{\beta} S_{t-1}/\lambda, \quad S_0 = \overline{P}_{\beta}, \tag{4.38}$$

$$\begin{aligned} V_t = {} & (I_l - S_{t-1}(C_t^{\beta})^T(\lambda I_m + C_t^{\beta} S_{t-1}(C_t^{\beta})^T)^{-1} C_t^{\beta}) V_{t-1} \\ & - S_{t-1}(C_t^{\beta})^T(\lambda I_m + C_t^{\beta} S_{t-1}(C_t^{\beta})^T)^{-1} C_t^{\alpha}, \quad V_0 = 0_{l \times r}, \end{aligned} \tag{4.39}$$

$$W_t = \lambda W_{t-1} + \lambda (C_t^{\beta} V_{t-1} + C_t^{\alpha})^T (\lambda I_m + C_t^{\beta} S_{t-1}(C_t^{\beta})^T)^{-1}(C_t^{\beta} V_{t-1} + C_t^{\alpha}), \quad W_0 = 0_{r \times r}, \tag{4.40}$$

$$C_t^{\beta} = C_t^{\beta}(x_{t-1}) = \partial[\Phi(z_t, \beta_{t-1})\alpha_{t-1}]/\partial \beta_{t-1}, \quad C_t^{\alpha} = C_t^{\alpha}(x_{t-1}) = \Phi(z_t, \beta_{t-1}). \tag{4.41}$$

The assertion of this theorem will follow from Lemmas 4.1 and 4.2 if we omit in Eq. (4.31) the terms starting with the first order of smallness $O(\mu^{-1})$.

This algorithm will be called a diffuse training algorithm (the DTA) for model Eq. (4.1) with separable structure.

Let us formulate some consequences of obtained results.

Consequence 4.1.
The diffuse component $P_t^{dif} = \tilde{V}_t(I_r - W_t W_t^+)\tilde{V}_t^T$ is the term in expansion P_t which is proportional to a large parameter and it is equal to zero for $t \geq tr$, where $tr = \min_t\{t: W_t > 0, t = 1, 2, \ldots, N\}$.

Consequence 4.2.
The matrix K_t^{dif} does not depend on the diffuse component as opposed to the matrix P_t and as the function μ is uniformly bounded in $t \in T$ as $\mu \to \infty$.

Consequence 4.3.
Numerical implementation errors can result in the GN method divergence for large values of μ. Indeed, let δW_t^+ be the error connected with calculations of the pseudoinverse W_t. Then by Lemma 4.2

$$K_t = (\tilde{S}_t + \tilde{V}_t L_t^{-1} \tilde{V}_t^T) C_t^T$$
$$= [\tilde{S}_t + \tilde{V}_t (I_r - W_t (W_t^+ + \delta W_t^+)\mu + O(1)) \tilde{V}_t^T] C_t^T, \quad t = 1, 2, \ldots, N,$$
$$\mu \to \infty.$$

For $\delta W_t^+ \neq 0$ the matrix K_t becomes dependent on the diffuse component. Moreover, this implies that even when $t \geq tr$ the passage to the use of the representation

$$P_t = (P_{t-1} - P_{t-1} C_t^T (\lambda I_m + C_t P_{t-1} C_t^T)^{-1} C_t P_{t-1}) / \lambda$$

can turn out to be unjustified since the matrix P_{tr} can still be ill-conditioned.

Numerically implemented, the DTA does not have the mentioned distinctive features. This is evidenced by the absence of diffuse components, i.e., quantities proportional to the large parameter in its construction.

We also note one more important advantage of the DTA compared to the GN method with the soft initialization, namely, the selection of the value of μ is not required.

Now we will derive the conditions under which the estimate $x = (\alpha^T, \beta^T)^T$, obtained by means of the GN method with the bounded value of μ, asymptomatically approaches x_t^{dif} as $\mu \to \infty$.

Let us consider a more compact representation form of the training algorithms. We introduce the vector u_t, whose elements are ordered by the column elements of x_t, S_t, V_t, and L_t and which are defined by the equations system

$$u_t = f_t(u_{t-1}, 1/\mu), \quad t \in T, \tag{4.42}$$

with the initial condition $u_0 = \overline{u}$. By means of these notations the DTA can be described by the system of equations

$$u_t^{dif} = f_t(u_{t-1}^{dif}, 0), \quad t \in T, \tag{4.43}$$

with the initial condition $u_0 = \overline{u}$, where u_t^{dif} includes ordered by the columns elements of x_t^{dif}, S_t^{dif}, V_t^{dif}, and W_t^{dif}.

Theorem 4.2.

Let the conditions be satisfied:

1. $P(||\xi_t|| < \infty) = 1$, $t \in T = \{1, 2, \ldots, N\}$, where T is bounded.
2. The solution of the system Eq. (4.43) belongs to the area $\tilde{U} = \{t \in T, ||u_t^{dif}|| \leq h\}$, where $h > 0$ is some number.
3. $||\Phi(z_t, \beta)|| \leq k_1$, $||\partial\Phi(z_t, \beta)/\partial\beta|| \leq k_2$, $||\partial^2\Phi(z_t, \beta)/\partial\beta_i\partial\beta_j|| \leq k_3$, $i, j = 1, 2, \ldots, l$ for $t \in T$ and $||\beta|| \leq h$, where k_1, k_2, and k_3 are some positive numbers.
4. $rank(W_t)$ is constant for every fixed $t \in T$, $||W_t|| \leq h$ and $\mu > \overline{\mu}$, where $\overline{\mu}$ is some sufficiently large number.

Then with probability one in the area $U = \{t \in T, ||u_t|| \leq h, \mu > \overline{\mu}\}$ the uniform asymptotic representation in $t \in T$ is true

$$x_t = x_t^{dif} + O(1/\mu), \quad \mu \to \infty, \quad t \in T, \tag{4.44}$$

where

$$x_t^{dif} = x_{t-1}^{dif} + K_t^{dif}(y_t - h_t(x_{t-1}^{dif})), \quad x_0^{diff} = (\overline{\beta}^T, 0_{1\times r})^T. \tag{4.45}$$

Proof

Introducing a new variable $e_t = u_t - u_t^{dif}$, we obtain

$$\begin{aligned} e_t &= e_{t-1} + f_t(u_{t-1}, 1/\mu) - f_t(u_{t-1}^{dif}, 0) \\ &= e_{t-1} + f_t(e_{t-1} + u_{t-1}^{dif}, 1/\mu) - f_t(u_{t-1}^{dif}, 0) \\ &= e_{t-1} + [f_t(e_{t-1} + u_{t-1}^{dif}, 1/\mu) - f_t(u_{t-1}^{dif}, 1/\mu)] + [f_t(u_{t-1}^{dif}, 1/\mu) - f_t(u_{t-1}^{dif}, 0)] \\ &= e_{t-1} + \Psi_t(e_{t-1}, 1/\mu) + \varphi_t(u_{t-1}^{dif}, 1/\mu), \quad e_0 = 0, \quad t \in T, \end{aligned}$$

or equivalently

$$e_t = \sum_{i=0}^{t-1} \Psi_i(e_i, 1/\mu) + \tilde{\varphi}_t(1/\mu), \quad e_0 = 0, \quad t \in T, \tag{4.46}$$

where $\tilde{\varphi}_t(1/\mu) = \sum_{i=0}^{t-1} \varphi_i(u_i^{dif}, 1/\mu)$.

Using the mean-value theorem in $\tilde{U}$, gives

$$f_t(e_{t-1} + u_{t-1}^{dif}, 1/\mu) - f_t(u_{t-1}^{dif}, 1/\mu) = \int_0^1 \frac{\partial f_t(z, 1/\mu)}{\partial z}\Big|_{z=u_{t-1}^{dif}+\tau e_{t-1}} d\tau \; e_{t-1}, \tag{4.47}$$

$$f_t(u_{t-1}^{dif}, 1/\mu) - f_t(u_{t-1}^{dif}, 0) = \int_0^1 \frac{\partial f_t(u_{t-1}^{dif}, z)}{\partial z}\Big|_{z=\tau/\mu} d\tau/\mu \tag{4.48}$$

so long as derivatives exist in these expressions.

It follows from the right-hand sides of Eqs. (4.12) and (4.13) that under the conditions 1 and 2 of the theorem their partial derivatives with respect to u_t are bounded in U. The existence and the boundedness of the partial derivatives of the functions in the right-hand part of the systems Eqs. (4.6) and (4.45) depend on the matrix function properties

$$K_t = P_t C_t^T R_t^{-1} = (\tilde{S}_t + \tilde{V}_t L_t \tilde{V}_t^T) C_t^T, \tag{4.49}$$

$$K_t^{dif} = (\tilde{S}_t^{dif} + \tilde{V}_t^{dif} (W_t^{dif})^+ (\tilde{V}_t^{dif})^T) C_t^T. \tag{4.50}$$

We have

$$\begin{aligned}L_t^{-1} &= \lambda^t \left(\sum_{k=1}^{t} \lambda^{-k+1} (C_t \tilde{V}_t)^T (\lambda I_m + C_t^\beta S_{t-1} (C_t^\beta)^T)^{-1} C_t \tilde{V}_t + \overline{P}_\alpha^{-1} / \mu \right) \\ &= \lambda^t \overline{P}_\alpha^{-1/2} \left(\sum_{k=1}^{t} \lambda^{-k+1} \overline{P}_\alpha^{1/2} (C_t \tilde{V}_t)^T (\lambda I_m + C_t^\beta S_{t-1} (C_t^\beta)^T)^{-1} C_t \tilde{V}_t \overline{P}_\alpha^{1/2} + I_r / \mu \right) \overline{P}_\alpha^{-1/2} \\ &= \lambda^t \overline{P}_\alpha^{-1/2} (\Lambda_t + I_r / \mu) \overline{P}_\alpha^{-1/2}.\end{aligned}$$

Using the identity [47]

$$(I_n + \mu V)^{-1} = (I_n - VV^+) + V^+ (V^+ + \mu I_n)^{-1},$$

where V is an arbitrary symmetric matrix and assuming $V = \Lambda_t$, we obtain

$$\begin{aligned}L_t &= \lambda^{-t} \overline{P}_\alpha^{1/2} (I_r + \mu \Lambda_t)^{-1} \overline{P}_\alpha^{1/2} \mu \\ &= \lambda^{-t} \overline{P}_\alpha (I_r - \Lambda_t \Lambda_t^+) \mu + \lambda^{-t} \overline{P}_\alpha^{1/2} \Lambda_t^+ (\Lambda_t^+ + \mu I_r)^{-1} \overline{P}_\alpha^{1/2} \mu.\end{aligned} \tag{4.51}$$

We have by Eqs. (4.34) and (4.51)

$$\begin{aligned}K_t &= \tilde{S}_t C_t^T + \lambda^{-t} \tilde{V}_t \overline{P}_\alpha^{1/2} (I_r + \mu \Lambda_t)^{-1} \overline{P}_\alpha^{1/2} \tilde{V}_t^T C_t^T \mu \\ &= \tilde{S}_t C_t^T + \overline{P}_\alpha^{1/2} W_t^+ (\lambda^t \overline{P}_\alpha^{-1/2} W_t^+ \overline{P}_\alpha^{-1/2} / \mu + I_r)^{-1} \overline{P}_\alpha^{1/2} C_t^T.\end{aligned} \tag{4.52}$$

If condition 4 of the theorem is satisfied then the matrix function $W_t^+(W_t)$is continuous and has a continuous derivative with respect to W_t in $U_w = \{t \in T, ||W_t|| \le h\}$ [1] and therefore $||W_t^+||$ is bounded in this area. Since

$$\begin{aligned}&||(I_r + \lambda^t \overline{P}_\alpha^{-1/2} W_t^+ \overline{P}_\alpha^{-1/2} / \mu)^{-1}|| = ||T_t diag((1 + \lambda_t(1)/\mu)^{-1}, \\ &\quad (1 + \lambda_t(2)/\mu)^{-1}, \ldots, (1 + \lambda_t(r)/\mu)^{-1}) T_t^T|| \le \\ &\le ||T_t||^2 \left(\sum_{i=1}^{r} (1 + \lambda_t(i)/\mu)^{-2} \right)^{1/2} \le r ||T_t||^2, \quad \mu > \overline{\mu},\end{aligned}$$

where $\lambda_t(i)$, $T_t(i) t \in T, i = 1, 2, \ldots, r$ are eigenvalues and corresponding eigenvectors of the matrix $\lambda^t \overline{P}_\alpha^{-1/2} W_t^+ \overline{P}_\alpha^{-1/2}$, $T_t = (T_t(1), T_t(2), \ldots, T_t(r))$ is an orthogonal matrix then the norm of the matrix function

$$\Psi(W_t, \mu) = (I_r + \lambda^t \overline{P}_\alpha^{-1/2} W_t^+ \overline{P}_\alpha^{-1/2} / \mu)^{-1}$$

is bounded as well.

Differentials of $\Psi(W_t, \mu)$ with respect to W_t^+ and μ are defined by the expressions

$$d_{W^+}\Psi(W_t, \mu) = -\lambda^t (I_r + \lambda^t \overline{P}_\alpha^{-1/2} W_t^+ \overline{P}_\alpha^{-1/2} / \mu)^{-1} \overline{P}_\alpha^{-1/2} dW_t^+ \overline{P}_\alpha^{-1/2} \times (I_r + \lambda^t \overline{P}_\alpha^{-1/2} W_t^+ \overline{P}_\alpha^{-1/2} / \mu)^{-1} / \mu,$$

$$d_\mu \Psi(W_t, \mu) = \lambda^t (I_r + \lambda^t \overline{P}_\alpha^{-1/2} W_t^+ \overline{P}_\alpha^{-1/2} / \mu)^{-1} \overline{P}_\alpha^{-1/2} W_t^+ \overline{P}_\alpha^{-1/2} \times (I_r + \lambda^t \overline{P}_\alpha^{-1/2} W_t^+ \overline{P}_\alpha^{-1/2} / \mu)^{-1} / \mu^2 d\mu$$

which are continuous functions of W_t and μ for $t \in T$, $\mu > \overline{\mu}$, and $||W_t|| \leq h$. Thus the derivatives in Eqs. (4.47) and (4.48) exist and are bounded in $\tilde{U}$ and therefore there are $c_1 > 0$, $c_2 > 0$ such that

$$||\psi_t(e_{t-1}, 1/\mu)|| = ||f_t(e_{t-1} + u_{t-1}^{dif}, 1/\mu) - f_t(u_{t-1}^{dif}, 1/\mu)|| \leq c_1 \left\| e_{t-1} \right\|, \tag{4.53}$$

$$||\varphi_t(u_{t-1}^{dif}, 1/\mu)|| = ||f_t(u_{t-1}^{dif}, 1/\mu) - f_t(u_{t-1}^{dif}, 0)|| \leq c_2/\mu. \tag{4.54}$$

Iterating Eq. (4.46), we obtain

$$||e_1|| \leq b_1/\mu, \quad ||e_2|| \leq c_1^2 ||e_1|| + b_2/\mu \leq (b_1 c_1^2 + b_2)/\mu, \quad \ldots,$$

$$||e_t|| \leq c_1^t ||e_1|| + c_2^t ||e_2|| + \cdots + c_{t-1}^t ||e_{t-1}|| + b_t/\mu = O(1/\mu),$$

where c_i^t, $i = 1, 2, \ldots, t-1$, b_i, $i = 1, 2, \ldots, t$ are some positive numbers.

Since $u_t^{dif} \in \tilde{U}$ then it follows from this that $u_t = u_t^{dif} + e_t \in U$ for $\mu > \overline{\mu}$ and at the same time we have the asymptotic representation Eq. (4.44).

Let us assume now that in the quality criteria instead of the forgetting factor the covariance matrix of the observation noise R_t is used

$$J_t(x) = \sum_{k=1}^{t} (y_k - \Phi(z_k, \beta)\alpha)^T R_k^{-1} (y_k - \Phi(z_k, \beta)\alpha) + \Gamma(\beta) + \alpha \overline{P}_\alpha^{-1} \alpha / \mu, \quad t = 1, 2, \ldots, N. \tag{4.55}$$

The solution in this case is not difficult to obtain, using Theorem 4.1. Indeed, by replacing

$$\tilde{y}_k = R_k^{-1/2} y_k, \quad \tilde{\Phi}(z_k, \beta) = R_k^{-1/2} \Phi(z_k, \beta) \tag{4.56}$$

Eq. (4.55) is transformed to Eq. (4.4) with $\lambda = 1$

$$\begin{aligned} J_t(x) = &\sum_{k=1}^{t} (\tilde{y}_k - \tilde{\Phi}(z_k, \beta)\alpha)^T (\tilde{y}_k - \tilde{\Phi}(z_k, \beta)\alpha) \\ &+ \Gamma(\beta) + \alpha \overline{P}_\alpha^{-1} \alpha/\mu, \quad t = 1, 2, \ldots, N. \end{aligned} \tag{4.57}$$

It follows from this the assertion:

Theorem 4.3.
The solution of the minimization problem (4.55) by the GN method with the diffuse initialization can be found recursively as follows

$$x_t^{dif} = x_{t-1}^{dif} + K_t^{dif}(y_t - h_t(x_{t-1}^{dif})), \quad x_0^{dif} = (\overline{\beta}^T, 0_{1\times r})^T, \quad t = 1, 2, \ldots, N, \tag{4.58}$$

where

$$x_t^{dif} = ((\beta_t^{dif})^T, (\beta_t^{dif})^T)^T, \; K_t^{dif} = ((K_t^\beta)^T, (K_t^\alpha)^T)^T,$$
$$h_t(x_{t-1}) = \Phi(z_t, \beta_{t-1})\alpha_{t-1}$$

$$K_t^\beta = (S_t + V_t W_t^+ V_t^T)(C_t^\beta)^T R_t^{-1} + V_t W_t^+ (C_t^\alpha)^T R_t^{-1}, \tag{4.59}$$

$$K_t^\alpha = W_t^+ V_t^T (C_t^\beta)^T R_t^{-1} + W_t^+ (C_t^\alpha)^T R_t^{-1}, \tag{4.60}$$

$$S_t = S_{t-1} - S_{t-1}(C_t^\beta)^T (R_t + C_t^\beta S_{t-1}(C_t^\beta)^T)^{-1} C_t^\beta S_{t-1}, \quad S_0 = \overline{P}_\beta, \tag{4.61}$$

$$\begin{aligned} V_t = &(I_l - S_{t-1}(C_t^\beta)^T (R_t + C_t^\beta S_{t-1}(C_t^\beta)^T)^{-1} C_t^\beta) V_{t-1} \\ &- S_{t-1}(C_t^\beta)^T (R_t + C_t^\beta S_{t-1}(C_t^\beta)^T)^{-1} C_t^\alpha, \quad V_0 = 0_{l\times r}, \end{aligned} \tag{4.62}$$

$$\begin{aligned} &W_t = W_{t-1} + (C_t^\beta V_{t-1} + C_t^\alpha)^T (R_t + C_t^\beta S_{t-1}(C_t^\beta)^T)^{-1} (C_t^\beta V_{t-1} + C_t^\alpha), \\ &W_0 = 0_{r\times r}, \end{aligned} \tag{4.63}$$

$$C_t^\beta = C_t^\beta(x_{t-1}) = \partial[\Phi(z_t, \beta_{t-1})\alpha_{t-1}]/\partial\beta_{t-1}, \quad C_t^\alpha = C_t^\alpha(x_{t-1}) = \Phi(z_t, \beta_{t-1}). \tag{4.64}$$

A successful choice of a priori information for β allows one to expect the fast convergence of the DTA to one of acceptable minimum points of the criterion. We first consider the following limiting case: $\overline{P}_\beta = 0_{r\times r}$. It follows from Theorem 4.1 that in Eq. (4.1) only the parameter α is estimated by the relations

$$\alpha_t^{dif} = \alpha_{t-1}^{dif} + K_t^{dif}(y_t - C_t^\alpha \alpha_{t-1}^{dif}),\quad \alpha_0^{dif} = 0_{r\times 1},\quad t = 1, 2, \ldots, N, \quad (4.65)$$

where

$$K_t^{dif} = W_t^+ (C_t^\alpha)^T, \quad (4.66)$$

$$W_t = \lambda W_{t-1} + (C_t^\alpha)^T C_t^\alpha,\quad W_0 = 0_{r\times r}. \quad (4.67)$$

If $\xi_t = 0$ then the equations system for the estimation error $e_t = \alpha_t - \alpha$ has the form

$$e_t = (I_r - K_t C_t^\alpha) e_{t-1} = A_t e_t,\quad e_0 = -\alpha,\quad t = 1, 2, \ldots, N$$

and its transition matrix is defined by the expression $H_{t,0} = I_r - W_t^+ W_t$ (Lemma 2.4). Assume that there is $tr = \min_t\{t{:}\Pi_t > 0, t = 1, 2, \ldots, N\}$, where $\Pi_t = \sum_{k=1}^t (C_k^\alpha)^T C_k^\alpha$. Then the vector estimation error e_t vanishes when $t \geq tr$. If $\xi_t \neq 0$ then it means that the DTA estimate will be unbiased when $t \geq tr$.

The considered limiting case illustrates one more important distinctive feature of the DTA, namely, the initial stage of training ($t \leq tr$) can make an essential impact on the number of iterations necessary for the convergence of a nonlinear training problem. The training process with the help of the DTA looks like fast and slow movements that occur in dynamical systems. In our case β_t^{dif} is a slow variable and α_t^{dif} is fast. Because of this, the point $((\alpha_t^{dif})^T, (\beta_t^{dif})^T)^T$ reaches a neighborhood of an acceptable minimum for a finite number of steps $t \geq tr$. This property of the DTA is considered in more detail in Section 4.4.2.

The situation in which values of the norm $\overline{P}_\beta$ are different from zero but comparatively small seems to be more interesting. In this case different variants of construction of simplified treaning algorithms are possible. Let, for example, $\overline{P}_\beta = O(\varepsilon)$, $\varepsilon \to 0$. Then $S_t = O(\varepsilon)$, $V_t = O(\varepsilon)$, $\varepsilon \to 0$ uniformly in t for bounded set T. Neglecting by terms $V_t W_t^+ V_t^T (C_t^\beta)^T$, $V_t W_t^+ (C_t^\alpha)^T W_t^+ V_t^T (C_t^\beta)^T$ in Eqs. (4.36) and (4.37) and $C_t^\beta S_{t-1} (C_t^\beta)^T$, $C_t^\beta V_{t-1}$ in Eq. (4.40), we obtain a simplified version of the DTA from Theorem 4.1

$$K_t^\beta = S_t(C_t^\beta)^T, \quad K_t^\alpha = W_t^+(C_t^\alpha)^T, \tag{4.68}$$

$$S_t = S_{t-1}/\lambda - S_{t-1}(C_t^\beta)^T(\lambda I_m + C_t^\beta S_{t-1}(C_t^\beta)^T)^{-1} C_t^\beta S_{t-1}/\lambda, \quad S_0 = \overline{P}_\beta, \tag{4.69}$$

$$W_t = \lambda W_{t-1} + (C_t^\alpha)^T C_t^\alpha, \quad W_0 = 0_{r\times r}. \tag{4.70}$$

Similarly, for the DTA from Theorem 4.3 we get

$$K_t^\beta = S_t(C_t^\beta)^T R_t^{-1}, \quad K_t^\alpha = W_t^+(C_t^\alpha)^T R_t^{-1}, \tag{4.71}$$

$$S_t = S_{t-1} - S_{t-1}(C_t^\beta)^T(R_t + C_t^\beta S_{t-1}(C_t^\beta)^T)^{-1} C_t^\beta S_{t-1}, \quad S_0 = \overline{P}_\beta, \tag{4.72}$$

$$W_t = W_{t-1} + (C_t^\alpha)^T R_t^{-1} C_t^\alpha, \quad W_0 = 0_{r\times r}. \tag{4.73}$$

We can come to the same relations in a different way. Indeed, solving the optimization problem and alternately fixing parameters, we obtain relations (4.68)–(4.73). This implies the name a two-stage training algorithm with the diffuse initialization.

4.3 TRAINING IN THE ABSENCE OF A PRIORI INFORMATION ABOUT PARAMETERS OF THE OUTPUT LAYER

Let us assume now that the unknown parameters in Eq. (4.1) satisfy conditions B1 and B2 and the quality criteria at a moment t is defined by Eq. (4.5). We will need some auxiliary statements to obtain the main result of this section, namely, Theorem 4.4.

Lemma 4.3.
The following identity is true:

$$\Lambda_t^+ \tilde{C}_t^T = [(I_{l+r} - \lambda^{-t/2} K_t C_t)\Lambda_{t-1}^+ \tilde{C}_{t-1}^T, K_t], \quad t = 1, 2, \ldots, N, \tag{4.74}$$

where

$$C_t \in R^{m\times(l+r)}, \; \tilde{C}_t = (\lambda^{-1/2} C_1^T, \lambda^{-1} C_2^T, \ldots, \lambda^{-t/2} C_t^T)^T \in R^{mt\times(l+r)},$$

$$\Lambda_t = (\tilde{C}_t^T \tilde{C}_t + M) \in R^{(l+r)\times(l+r)},$$

$$K_t = \lambda^{-t/2}\Lambda_t^+ C_t^T \in R^{(l+r)\times m}, M = \begin{pmatrix} \overline{P}_\beta^{-1} & 0_{l\times r} \\ 0_{r\times l} & 0_{r\times r} \end{pmatrix} \in R^{(l+r)\times(l+r)}, \overline{P}_\beta \in R^{l\times l},$$

$0_{p\times q}$ is a $p \times q$ matrix with zero elements.

Proof

Since $\Lambda_t^+ \tilde{C}_t^T = \Lambda_t^+(\tilde{C}_{t-1}^T, \lambda^{-t/2}C_t^T)^T$ then the last m columns in the left- and right-hand sides of identity Eq. (4.74) coincide. Let us show that

$$\Lambda_t^+ \tilde{C}_{t-1}^T = (I_{l+r} - \lambda^{-t/2}K_t C_t)\Lambda_{t-1}^+ \tilde{C}_{t-1}^T.$$

Using the recursive formula

$$\Lambda_t = \Lambda_{t-1} + \lambda^{-t} C_t^T C_t, \quad \Lambda_0 = M, \quad t = 1, 2, \ldots, N$$

for the matrix determination of Λ_t, we represent this expression in the equivalent form

$$\begin{aligned}
&(\Lambda_t^+ - (I_{l+r} - \lambda^{-t/2}K_t C_t)\Lambda_{t-1}^+)\tilde{C}_{t-1}^T \\
&= (\Lambda_t^+ - (I_{r+l} - \Lambda_t^+\Lambda_t + \Lambda_t^+\Lambda_{t-1})\Lambda_{t-1}^+)\tilde{C}_{t-1}^T \\
&= (\Lambda_t^+(I_{r+l} - \Lambda_{t-1}\Lambda_{t-1}^+) - (I_{r+l} - \Lambda_t^+\Lambda_t)\Lambda_{t-1}^+)\tilde{C}_{t-1}^T \\
&= 0.
\end{aligned} \tag{4.75}$$

Let us show that

$$(I_{l+r} - \Lambda_t^+\Lambda_t)\Lambda_{t-1}^+ = 0, \tag{4.76}$$

$$\Lambda_t^+(I_{l+r} - \Lambda_{t-1}\Lambda_{t-1}^+)\tilde{C}_{t-1}^T = 0. \tag{4.77}$$

Let the matrix $Q_t = (M^{1/2}, \tilde{C}_t^T)$ have a rank $q(t)$ and $l_t(1), l_t(2), \ldots, l_t(q(t))$ be all its arbitrary linearly independed columns and $L_t = (l_t(1), l_t(2), \ldots, l_t(q(t))$. The matrix L_i is selected so that $L_i = (L_{i-1}, \Delta_i)$, where Δ_i is matrix of rank $q(i) - q(i-1)$, composed of all linearly independent columns of the matrix $\lambda^{-i/2}C_i^T$ for each $i = 2, 3, \ldots, t$. The linear space $C(Q_t)$ defined by them coincides with the space formed by columns Λ_t which provide $C(\Lambda_{t-1}) \subseteq C(\Lambda_t)$.

Using the skeleton decomposition of the matrices, we get

$$(\tilde{C}_t^T, M^{1/2}) = L_t\Gamma_t,$$

where $rank(\Gamma_t) = q(t)$.

We have $\Lambda_t = L_t\tilde{\Gamma}_t L_t^T$, where $\tilde{\Gamma}_t = \Gamma_t\Gamma_t^T$. Since L_t is a matrix of full rank with respect to columns, we have $L_t^+ = (L_t^T L_t)^{-1} L_t^T$ and

$$\Lambda_t^+ = (L_t\tilde{\Gamma}_t L_t^T)^+ = (L_t^T)^+\tilde{\Gamma}_t^{-1}(L_t)^+, \quad \Lambda_t^+\Lambda_t = L_t(L_t^T L_t)^{-1}L_t^T,$$

$$(I_{l+r} - L_t(L_t^T L_t)^{-1}L_t^T)L_{t-1} = 0$$

that implies Eq. (4.76).

Consider the equality Eq. (4.77). Since there is a matrix $\overline{\Gamma}_t$ such that $\tilde{C}_{t-1}^T = L_{t-1}\overline{\Gamma}_t$, we obtain

$$\Lambda_t^+(I_{l+r} - \Lambda_{t-1}\Lambda_{t-1}^+)\tilde{C}_{t-1}^T = \Lambda_t^+(I_{l+r} - L_{t-1}(L_{t-1}^T L_{t-1})^{-1}L_{t-1}^T)L_{t-1}\overline{\Gamma}_t = 0,$$
$$t = 1, 2, \ldots, N.$$

Lemma 4.4.

There is uniform in respect to $t \in T = \{1, 2, \ldots, N\}$ the asymptotic representation

$$(\varepsilon U_t + M_t)^{-1} = 1/\varepsilon A_{-1} + A_0 + O(\varepsilon), \quad \varepsilon \to 0, \quad t = 1, 2, \ldots, N, \tag{4.78}$$

where

$$A_0 = M_t^+ = (\tilde{C}_t^T\tilde{C}_t + \lambda^t M)^+, \; U_t = \begin{pmatrix} 0_{l\times l} & 0_{l\times r} \\ 0_{r\times l} & \lambda^t I_r \end{pmatrix} \in R^{(l+r)\times(l+r)},$$

$$M = \begin{pmatrix} \overline{P}_\beta^{-1} & 0_{l\times r} \\ 0_{r\times l} & 0_{r\times r} \end{pmatrix} \in R^{(l+r)\times(l+r)}, \quad \overline{P}_\beta \in R^{l\times l},$$

$C_t \in R^{m\times(l+r)}$, $\tilde{C}_t = (\lambda^{(t-1)/2}C_1^T, \lambda^{(t-2)/2}C_2^T, \ldots, C_t^T)^T \in R^{mt\times(l+r)}$, $\varepsilon > 0$ is a small parameter, T is a bounded set.

Proof

The existence of the asymptotic representation Eq. (4.78) follows from the expression [47]

$$(H^T H + 1/\varepsilon G^T G)^+ = (\overline{H}H)^+ + \varepsilon(I - \overline{H}^+ H)(G^T G)^+(I - \overline{H}^+ H)^T + O(\varepsilon^2),$$

where H and G are arbitrary matrices of corresponding dimensions,

$$\overline{H} = H(I - G^+ G) = H(I - (G^T G)^+ G^T G),$$

provided that $H = U_t^{1/2}$, $G = M_t^{1/2}$.

We need to show that

$$M_t^+ = (I_{l+r} - \overline{H}_t^+ U_t^{1/2})M_t^+(I_{l+r} - \overline{H}_t^+ U_t^{1/2})^T,$$

where $\overline{H}_t = U_t^{1/2}(I_{l+r} - M_t^+ M_t)$ or

$$\overline{H}_t^+ U_t^{1/2} M_t^+ = 0.$$

Let us at first show that

$$\overline{H}_t^+ = \lambda^{-t}(I_{l+r} - M_t^+ M_t)\overline{U},$$

where

$$\overline{U} = \begin{pmatrix} 0_{l\times l} & 0_{l\times r} \\ 0_{r\times l} & I_r \end{pmatrix}.$$

We use the following result [47]. Let A and B be arbitrary matrices. Then $(AB)^+ = B^+A^+$ if and only if they carry out the conditions

$$A^+ABB^TA^T = BB^TA^T,$$

$$BB^+A^TAB = A^TAB.$$

Putting $A = \overline{U}$, $B = I_{l+r} - M_t^+ M_t$, gives

$$(\overline{U} - I_{l+r})(I_{l+r} - M_t^+ M_t)\overline{U} = 0, \tag{4.79}$$

$$M_t^+ M_t \overline{U}(I_{l+r} - M_t^+ M_t) = 0. \tag{4.80}$$

Let $T_t = (t_t(1), t_t(2), \ldots, t_t(l+r))$ be an orthogonal matrix such that $M_t = T_t\Lambda_t T_t^T$ where $\Lambda_t = diag(\lambda_t(1), \lambda_t(2), \ldots, \lambda_t(l+r))$, $\lambda_t(i)$, $i = 1, 2, \ldots, l+r$ are eigenvalues of the matrix M_t.

Define the structure of the matrix T_t. We have

$$rank(M_t) = rank\left(\lambda^t M + \begin{pmatrix} (\tilde{C}_t^{\beta})^T\tilde{C}_t^{\beta} & (\tilde{C}_t^{\beta})^T\tilde{C}_t^{\alpha} \\ (\tilde{C}_t^{\alpha})^T\tilde{C}_t^{\beta} & (\tilde{C}_t^{\alpha})^T\tilde{C}_t^{\alpha} \end{pmatrix}\right)$$
$$= rank(\lambda^t M + (\tilde{C}_t^{\beta})^T\tilde{C}_t^{\beta}) + rank(S_t),$$

where

$$S_t = (\tilde{C}_t^{\alpha})^T\tilde{C}_t^{\alpha} - (\tilde{C}_t^{\beta})^T\tilde{C}_t^{\alpha}(I_l + (\tilde{C}_t^{\beta})^T\tilde{C}_t^{\beta})^{-1}(\tilde{C}_t^{\alpha})^T\tilde{C}_t^{\beta},$$

$$\tilde{C}_t = (\tilde{C}_t^{\beta}, \tilde{C}_t^{\alpha}),\quad \tilde{C}_t^{\beta} \in R^{mt\times l},\quad \tilde{C}_t^{\alpha} \in R^{mt\times r}.$$

This implies that $q(t) = rank(M_t) \geq l$. If $rank(M_t) = l + r$, then the statement of the lemma is obviously true. Let $l \leq q(t) < l + r$. The eigenvectors of the matrix M_t (the columns of the matrix T_t) that correspond to zero eigenvalues are determined from the system $M_t x_t = 0$. Using the block representation $x_t = (x_t^T(1), x_t^T(2))^T$, we obtain

$$\begin{pmatrix} \lambda^t \overline{P}_\beta^{-1} + (\tilde{C}_t^\beta)^T \tilde{C}_t^\beta & (\tilde{C}_t^\beta)^T \tilde{C}_t^\alpha \\ (\tilde{C}_t^\alpha)^T \tilde{C}_t^\beta & (\tilde{C}_t^\alpha)^T \tilde{C}_t^\alpha \end{pmatrix} \begin{pmatrix} x_t(1) \\ x_t(2) \end{pmatrix} = 0$$

or equivalently

$$\lambda^t \overline{P}^{-1}{}_\beta^{-1} x_t(1) + (\tilde{C}_t^\beta)^T \tilde{C}_t x_t = 0, \quad (\tilde{C}_t^\alpha)^T \tilde{C}_t x_t = 0.$$

The solution of this system is defined by the expressions

$$x_t(1) = 0, \quad x_t(2) = (I_r - (\tilde{C}_t^\alpha)^+ \tilde{C}_t^\alpha) f,$$

where $f \in R^r$ is an arbitrary vector. From this without any loss of generality, we assume that

$$\Lambda_t = diag(\lambda_t(1), \lambda_t(2), \ldots, \lambda_t(q(t)), 0, 0, \ldots, 0), \quad \lambda_t(i) > 0, i = 1, 2, \ldots, q(t)$$

and obtain

$$T_t = \begin{pmatrix} T_{1t} & 0_{l \times (l+r-q(t))} \\ T_{2t} & T_{3t} \end{pmatrix}, \quad T_{1t} \in R^{l \times q(t)}, \quad T_{2t} \in R^{r \times q(t)}, \quad T_{3t} \in R^{r \times (l+r-q(t))}.$$

Transforming the left-hand sides of Eqs. (4.79) and (4.80) with the help of the spectral decomposition $M_t^+ = T_t \Lambda_t^+ T_t^T$ gives

$$(\overline{U} - I_{l+r})(I_{l+r} - M_t^+ M_t)\overline{U} = (\overline{U} - I_{l+r}) T_t (I_{l+r} - \Lambda_t^+ \Lambda_t) T_t^T \overline{U},$$

$$M_t^+ M_t \overline{U}(I_{l+r} - M_t^+ M_t) = T_t \Lambda_t^+ \Lambda_t T_t^T \overline{U} T_t (I_{l+r} - \Lambda_t^+ \Lambda_t) T_t^T.$$

The view of the structure T_t and the relation $T_{3t}^T T_{2t} = 0_{(l+r-q(t)) \times q(t)}$ from here Eqs. (4.79), (4.80) follow.

Similarly, it is verified that

$$\overline{H}_t^+ U_t^{1/2} M_t^+ = (I_{l+r} - M_t^+ M_t)\overline{U} M_t^+ = T_t (I_{l+r} - \Lambda_t^+ \Lambda_t) T_t^T \overline{U} T_t \Lambda_t^+ T_t^T = 0.$$

Lemma 4.5.
The following representation is true:

$$M_t^+ = (\tilde{C}_t^T \tilde{C}_t + \lambda^t M)^+ = \tilde{S}_t + \tilde{V}_t W_t^+ \tilde{V}_t^T, \quad t = 1, 2, \ldots, N, \qquad (4.81)$$

where

$$M = \begin{pmatrix} \overline{P}_\beta^{-1} & 0_{l\times r} \\ 0_{r\times l} & 0_{r\times r} \end{pmatrix} \in R^{(l+r)\times(l+r)}, \quad \overline{P}_\beta \in R^{l\times l},$$

$$U_t = \begin{pmatrix} 0_{l\times l} & 0_{l\times r} \\ 0_{r\times l} & \lambda^t I_r \end{pmatrix} \in R^{(l+r)\times(l+r)},$$

$$\tilde{S}_t = \begin{pmatrix} S_t & 0_{l\times r} \\ 0_{r\times l} & 0_{r\times r} \end{pmatrix}, \quad \tilde{V}_t = \begin{pmatrix} V_t \\ I_r \end{pmatrix}, \quad S_t \in R^{l\times l}, \quad V_t \in R^{l\times r}, \tag{4.82}$$

S_t, V_t, W_t are defined by the matrix equations (4.38)–(4.40) for arbitrary $C_t^\beta \in R^{m\times l}$, $C_t^\alpha \in R^{m\times r}$.

Proof

The matrix

$$P_t^{-1} = \varepsilon U_t + M_t = \varepsilon \begin{pmatrix} 0_{l\times l} & 0_{l\times r} \\ 0_{r\times l} & \lambda^t I_r \end{pmatrix} + M_t$$

satisfies the matrix difference equation

$$P_t^{-1} = \lambda P_{t-1}^{-1} + C_t^T C_t, \quad P_0^{-1} = block\ diag(\overline{P}_\beta^{-1}, \varepsilon I_r), \quad t = 1, 2, \ldots, N,$$

where $C_t = (C_t^\beta, C_t^\alpha)$.

Let us use the matrix identity Eq. (2.7) to determine P_t. Putting in Eq. (2.7) $B = P_{t-1}, D = \lambda I_m$, $C = C_t^T$, we obtain

$$\begin{aligned} P_t &= (\lambda P_{t-1}^{-1} + C_t^T C_t)^{-1} = (P_{t-1} - P_{t-1} C_t^T (\lambda I_m + C_t P_{t-1} C_t^T)^{-1} C_t P_{t-1})/\lambda, \\ &= P_{t-1}/\lambda - 1/\lambda^2 P_{t-1} C_t^T (I_m + 1/\lambda C_t P_{t-1} C_t^T)^{-1} C_t P_{t-1}, \end{aligned} \tag{4.83}$$

$$P_0 = block\ diag(\overline{P}_\beta, I_r/\varepsilon), \quad t = 1, 2, \ldots, N.$$

It follows from Lemma 4.1 that $P_t = \tilde{S}_t + \tilde{V}_t L_t \tilde{V}_t^T$, where the matrices $\tilde{S}_t$, $\tilde{V}_t$, L_t are defined by relations (4.11)–(4.14). Using Lemma 2.1, we find from Eq. (4.14) that

$$L_t = 1/\varepsilon \lambda^{-t} \overline{P}_\alpha (I_r - W_t W_t^+) + W_t^+ + O(\varepsilon), \quad \varepsilon \to 0.$$

Since

$$\begin{aligned} P_t &= (\varepsilon U_t + M_t)^{-1} = Q_t + \tilde{S}_t = \tilde{V}_t L_t \tilde{V}_t^T + \tilde{S}_t \\ &= 1/\varepsilon \lambda^{-t} \tilde{V}_t \overline{P}_\alpha (I_r - W_t W_t^+) \tilde{V}_t^T + \tilde{V}_t W_t^+ \tilde{V}_t^T + \tilde{S}_t + O(\varepsilon), \quad \varepsilon \to 0 \end{aligned} \tag{4.84}$$

then Eq. (4.81) follows from Lemma 4.4.

Consider the auxiliary linear observation model Eq. (4.16) and the optimization problem that is connected with it

$$(\alpha^*, \beta^*) = \text{argmin} J_t(\alpha, \beta), \quad \alpha \in R^r, \quad \beta \in R^l, \tag{4.85}$$

where

$$J_t(x) = \sum_{k=1}^{t} \lambda^{t-k}(y_k - C_k^\beta \beta - C_k^\alpha \alpha)^T (y_k - C_k^\beta \beta - C_k^\alpha \alpha) + \lambda^t \Gamma(\beta),$$
$$t = 1, 2, \ldots, N, \tag{4.86}$$

$$y_t \in R^m, \quad C_t^\beta \in R^{m \times l}, \quad C_t^\alpha \in R^{m \times r}, \quad \xi_t \in R^m.$$

Lemma 4.6.
The solution of problems (4.85) and (4.86) can be represented in the recursive form

$$\beta_t = \beta_{t-1} + K_t^\beta (y_t - C_t^\beta \beta_{t-1} - C_t^\alpha \alpha_{t-1}), \quad \beta_0 = \overline{\beta}, \tag{4.87}$$

$$\alpha_t = \alpha_{t-1} + K_t^\alpha (y_t - C_t^\beta \beta_{t-1} - C_t^\alpha \alpha_{t-1}), \quad \alpha_0 = 0_{r \times 1}, \quad t = 1, 2, \ldots, N, \tag{4.88}$$

where

$$K_t^\beta = (S_t + V_t W_t^+ V_t^T)(C_t^\beta)^T + V_t W_t^+ (C_t^\alpha)^T, \tag{4.89}$$

$$K_t^\alpha = W_t^+ V_t^T (C_t^\beta)^T + W_t^+ (C_t^\alpha)^T. \tag{4.90}$$

Proof
Assume that $x = (\alpha^T, \beta^T)^T$ and pass to a more compact form of representation of the criterion

$$J_t(x) = (Y_t - \tilde{C}_t x)^T (Y_t - \tilde{C}_t x) + \lambda^t (x - \overline{x})^T M (x - \overline{x}), \quad t = 1, 2, \ldots, N, \tag{4.91}$$

where $\overline{x} = (\overline{\beta}^T, 0_{1 \times r})^T$, the matrix M is defined in Lemma 4.3,

$$\tilde{C}_t = (\lambda^{(t-1)/2} C_1^T, \lambda^{(t-2)/2} C_2^T, \ldots, C_t^T)^T \in R^{mt \times (l+r)},$$

$$Y_t = (\lambda^{(t-1)/2} y_1^T, \lambda^{(t-2)/2} y_2^T, \ldots, y_t^T)^T \in R^{mt \times 1}.$$

Using the replacement $z = x - \bar{x}$, we obtain

$$J_t(z+\bar{x}) = ||\tilde{Y}_t - \tilde{C}_t z||^2 + \lambda^t z^T M z$$
$$= \left\| \begin{pmatrix} \tilde{Y}_t \\ 0_{(l+r)\times 1} \end{pmatrix} - \begin{pmatrix} \tilde{C}_t \\ \lambda^{t/2} M^{1/2} \end{pmatrix} z \right\|^2,$$

where $\tilde{Y}_t = Y_t - \tilde{C}_t \bar{x}$. Stationary points of this problem are determined by the system of the normal equations

$$\begin{pmatrix} \tilde{C}_t \\ \lambda^{t/2} M^{1/2} \end{pmatrix}^T \begin{pmatrix} \tilde{C}_t \\ \lambda^{t/2} M^{1/2} \end{pmatrix} z = \begin{pmatrix} \tilde{C}_t \\ \lambda^{t/2} M^{1/2} \end{pmatrix}^T \begin{pmatrix} \tilde{Y}_t \\ 0_{(l+r)\times 1} \end{pmatrix},$$

or with the use of the following equivalent form

$$(\tilde{C}_t^T \tilde{C}_t + \lambda^t M) z = \tilde{C}_t^T \tilde{Y}_t. \tag{4.92}$$

We will consider only the minimum norm solution which is defined by the expression

$$z^* = (\tilde{C}_t^T \tilde{C}_t + \lambda^t M)^+ \tilde{C}_t^T \tilde{Y}_t = M_t^+ \tilde{C}_t^T \tilde{Y}_t. \tag{4.93}$$

Let us show that it can be recursively found. Denote $z_t = z^*$. Using Lemma 4.3, gives

$$z_t = (\tilde{C}_t^T \tilde{C}_t + M)^+ \tilde{C}_t^T (\tilde{Y}_{t-1}^T, \tilde{y}_{t-1}^T)^T$$
$$= [(I_{l+r} - K_t C_t) M_{t-1}^+ \tilde{C}_{t-1}^T, K_t](\tilde{Y}_{t-1}^T, \tilde{y}_t^T)^T$$
$$= (I_{l+r} - K_t C_t) z_{t-1} + K_t \tilde{y}_t, \quad z_0 = 0, \quad t = 1, 2, \ldots, N,$$

where $K_t = M_t^+ C_t^T = (\tilde{C}_t^T \tilde{C}_t + \lambda^t M)^+ C_t^T$.

Coming back to the initial variable $x_t = z_t + \bar{x}$, we obtain

$$x_t = x_{t-1} + K_t(y_t - C_t x_{t-1}), \quad x_0 = (\bar{\beta}^T, 0_{1\times r})^T, \quad t = 1, 2, \ldots, N. \tag{4.94}$$

Expressions (4.89) and (4.90) follow from Lemma 4.5.

Now we can formulate an algorithm for the solution of the nonlinear minimization problem with criterion Eq. (4.5).

Theorem 4.4.

Let unknown parameters in Eq. (4.1) satisfy conditions B1 and B2. Then the solution to the problem of minimization of the criterion Eq. (4.5) by the GN method coincides with the DTA.

Proof

This assertion follows from Lemma 4.6.

Comparing the two problem statements of parameter estimation in Eq. (4.1), which were given in Section 4.1, we note the following. When solving the minimization problem with the quality criteria Eq. (4.5), any statistical assumptions about the nature of the unknown initial conditions are not used. It seems that this idea is more logically justified, unlike the diffuse initialization, in which we have to deal with infinite covariance matrix of the initial conditions. However from Theorem 4.4 it follows that these two settings lead to the same relations for parameter estimation.

4.4 CONVERGENCE OF DIFFUSE TRAINING ALGORITHMS

The DTA behavior with an increase of sample size is studied in this section. The problem specialty is related to the separable character of the observation models and the fact that the nonlinearly inputting parameters belong to some compact set, and linearly inputting parameters should be considered as arbitrary numbers.

We will suppose that the observations are generated by the model

$$y_t = \Phi(z_t, \beta^*)\alpha^* + \xi_t, \quad t = 1, 2, \ldots, \tag{4.95}$$

for some sequences $\xi_t \in R^m$, $z_t \in R^n$, where $\beta^* \in R^l$, $\alpha^* \in R^r$ are unknown parameters. It is assumed that for the estimation of the parameters in Eq. (4.95) the DTAs obtained in Section 4.2 are used and defined by the system of equations

$$x_t = x_{t-1} + K_t(y_t - h_t(x_{t-1})), \quad x_0 = (\overline{\beta}^T, 0_{1\times r})^T, \quad t = 1, 2, \ldots, \tag{4.96}$$

where $x_t = (\beta_t^T, \alpha_t^T)^T$, $h_t(x_{t-1}) = \Phi(z_t, \beta_{t-1})\alpha_{t-1}$ or in the equivalent form

$$\beta_t = \beta_{t-1} + K_t^{\beta}(y_t - h_t(x_{t-1})), \quad \beta_0 = \overline{\beta}, \tag{4.97}$$

$$\alpha_t = \alpha_{t-1} + K_t^{\alpha}(y_t - h_t(x_{t-1})), \quad \alpha_0 = 0_{r\times 1}, \quad t = 1, 2, \ldots, \tag{4.98}$$

where $K_t = ((K_t^{\beta})^T, (K_t^{\alpha})^T)^T$ is described by the expressions (4.36)−(4.41) or (4.59)−(4.64).

If $\xi_t = 0$ then the system Eq. (4.96) has the solution $x_t = x^* = ((\beta^*)^T, (\alpha^*)^T)^T$. The vector function $\eta_t = K_t\xi_t$ can be interpreted as disturbance acting on it. We want to find conditions under which $||x_t - x^*||$ becomes small with increasing t.

4.4.1 Finite Training Set

Along with the system Eq. (4.96) consider the system

$$z_t = z_{t-1} + K_t(h_t(x^*) - h_t(z_{t-1})), \quad z_0 = (\overline{\beta}^T, 0_{1\times r})^T, \quad t = 1, 2, \ldots \tag{4.99}$$

Definition 1

The solution $z_t = x^*$ of the system Eq. (4.99) is finitely stable under the action of the disturbance ξ_t if for any $\varepsilon > 0$, $d > 0$ there are such $\delta_1(A_d)$, $\delta_2(A_d)$, $\delta_3(A_d) > 0$ that $||\beta_0 - \beta^*|| < \delta_1(A_d)$, $\alpha_0 = 0$, $||\xi_t|| < \delta_2(A_d)$, $||S_0|| < \delta_3(A_d)$ imply $||x_t - x^*|| < \varepsilon$ for all $\alpha^* \in A_d = \{\alpha: ||\alpha|| \le d\} \subset R^r$, $t \ge tr = \min\{t: W_t > 0, t \in T\}$, where $t \in T = \{1, 2, \ldots, N\}$ is bounded set.

Note that this definition formalizes the idea inherent to the DTA of fast and slow movements.

We will need some auxiliary statements to obtain the main result of this section, namely, Theorem 4.5.

Lemma 4.7.

The solution of the equations system

$$x_t = A_t x_{t-1} + f_t, \quad x_0 = \overline{x}, \quad t = 1, 2, \ldots \tag{4.100}$$

is defined by the expressions

$$x_t = H_{t,0} x_0 + \sum_{s=1}^{t} H_{t,s} f_s, \quad t = 1, 2, \ldots, \tag{4.101}$$

where $x_t, f_t \in R^n$, $A_t \in R^{n\times n}$,

$$H_{t,s} = \begin{cases} A_t A_{t-1} \ldots A_{s+1}, & t = s+1, \ s+2, \ldots \\ I_n, & t = s \end{cases} \tag{4.102}$$

or in another form

$$H_{t,s} = A_t H_{t-1,s}, \quad t = s+1, \ s+2, \ldots, \quad H_{s,s} = I_n. \tag{4.103}$$

Proof

Iterating Eqs. (4.100) and (4.103) yields, respectively,

$$x_1 = A_1 x_0 + f_1, \quad x_2 = A_2 A_1 x_0 + A_2 f_1 + f_2, \quad \ldots,$$

$$x_t = A_t A_{t-1} \ldots A_1 x_0 + A_t A_{t-1} \ldots A_2 f_1 + \cdots + f_t,$$

$$H_{s+1,s} = A_{s+1},\ \ H_{s+2,s} = A_{s+2}A_{s+1},\ \ \ldots,\ \ H_{t,s} = A_tA_{t-1}\ldots A_{s+1}.$$

This implies Eqs. (4.101) and (4.102).

Lemma 4.8.

The solution of the equations system

$$Z_{t-1,s} = B_tZ_{t,s} + F_{t,s},\ \ Z_{s,s} = I_n,\quad t = s-1,\ s-2,\ldots \tag{4.104}$$

is defined by the expressions

$$Z_{t,s} = Y_{t,s} + \sum_{j=t}^{s-1} Y_{t,j}F_{j+1,s},\quad t = s-1, s-2,\ldots, \tag{4.105}$$

where $Z_{t,s}, F_{t,s}, A_t \in R^{n\times n}$,

$$Y_{t,s} = \begin{cases} B_{t+1}B_{t+2}\ldots B_s, & t = s-1,\ldots \\ I_n, & t = s \end{cases} \tag{4.106}$$

or in another form

$$Y_{t-1,s} = B_tY_{t,s},\ \ t = s-1,\ s-1,\ldots,\ \ Y_{s,s} = I_n. \tag{4.107}$$

Proof

Iterating Eqs. (4.104) and (4.107) yields, respectively,

$$Z_{s-1,s} = B_s + F_{s,s},\ \ Z_{s-2,s} = B_{s-1}B_s + B_{s-1}F_{s,s} + F_{s-1,s},\ \ \ldots,$$

$$Z_{t,s}=B_{t+1}B_{t+2}\ldots B_s + B_{t+1}B_{t+2}\ldots B_{s-1}F_{s,s} + B_{t+1}B_{t+2}\ldots B_{s-2}F_{s-1,s} + \cdots + F_{t+1,s},$$

$$Y_{s-1,s} = B_s,\ Y_{s-2,s} = B_{s-1}B_s,\ldots,\ Y_{t,s} = B_{t+1}B_{t+2}\ldots B_s.$$

$$Z_{t,s} = Y_{t,s} + Y_{t,s-1}F_{s,s} + Y_{t,s-2}F_{s-1,s} + \cdots + F_{t+1,s} = Y_{t,s} + \sum_{j=t}^{s-1} Y_{t,j}F_{j+1,s}.$$

This implies Eqs. (4.105) and (4.106).

We at first consider the convergence of the DTA from Theorem 4.3.

Let us represent expressions for the DTA from Theorem 4.3 in the following more compact and convenient form for further use

$$K_t = (\tilde{S}_t + \tilde{V}_t\tilde{W}_t^{+}\tilde{V}_t^{T})C_t^TR_t^{-1} = \tilde{S}_tC_t^TR_t^{-1} + K_t^d,\quad t = 1,2,\ldots, \tag{4.108}$$

where

$$\tilde{S}_t = \tilde{S}_{t-1} - \tilde{S}_{t-1} C_t^T N_t^{-1} C_t \tilde{S}_{t-1}, \tag{4.109}$$

$$\tilde{V}_t = (I_{l+r} - \tilde{S}_{t-1} C_t^T N_t^{-1} C_t) \tilde{V}_{t-1} = A_{1t} \tilde{V}_{t-1}, \tag{4.110}$$

$$\tilde{W}_t = \tilde{W}_{t-1} + \tilde{V}_{t-1}^T C_t^T N_t^{-1} C_t \tilde{V}_{t-1}, \tag{4.111}$$

$$N_t = R_t + C_t \tilde{S}_{t-1} C_t^T, \tag{4.112}$$

$$\tilde{S}_t = block\ diag(S_t, 0_{r\times r}),\ \ \tilde{V}_t = (V_t^T, I_r)^T,\ \ \tilde{W}_t = W_t,\ \ C_t = (C_t^\beta, C_t^\alpha). \tag{4.113}$$

Lemma 4.9.
Suppose that there is $tr = \min\{t: W_t > 0, t \in T\}$, $T = \{1, 2, \ldots, N\}$ and let $S_0 > 0$. Then for $t \geq tr$, $t \in T$ the following representations are true

$$H_{t,s} = \Phi_{t,s} - \tilde{V}_t W_t^{-1} \sum_{i=s}^{t-1} \tilde{V}_i^T C_{i+1}^T N_{i+1}^{-1} C_{i+1} \Phi_{i,s}, \tag{4.114}$$

$$H_{t,s} K_s^d = \tilde{V}_t W_t^{-1} \tilde{V}_s^T C_s^T R_s^{-1}, \quad t = s+1,\ s+2, \ldots, \tag{4.115}$$

where $H_{t,s}$, $\Phi_{t,s}$ are the solutions of the matrices equations systems

$$H_{t,s} = (I_{l+r} - K_t C_t) H_{t-1,s} = A_t H_{t-1,s}, \quad H_{s,s} = I_{l+r}, \tag{4.116}$$

$$\Phi_{t,s} = (I_{l+r} - \tilde{S}_t C_t^T R_t^{-1} C_t) \Phi_{t-1,s} = \tilde{A}_t \Phi_{t-1,s}, \quad \Phi_{s,s} = I_{l+r}, \tag{4.117}$$

K_t, $\tilde{V}_t$, W_t, $\tilde{S}_t$, N_t are defined by the expressions (4.108)–(4.113).

Proof
Consider two auxiliary systems of equations

$$G_{t-1,s} = A_t^T G_{t,s}, \quad G_{s,s} = I_{l+r}, \quad t = s-1, s-2, \ldots, \tag{4.118}$$

$$\begin{aligned} Z_{t-1,s} &= \tilde{A}_t^T Z_{t,s} - C_t^T N_t^{-1} C_t \tilde{V}_{t-1} W_s^{-1} \tilde{V}_s^T, \\ Z_{s,s} &= I_{l+r}, \quad t = s-1, s-2, \ldots, \quad s \geq tr. \end{aligned} \tag{4.119}$$

Let us first show that

$$G_{t,s} = Z_{t,s}, \quad t = s-1, s-2, \ldots, \quad s \geq tr. \tag{4.120}$$

Putting in Eq. (4.104)

$$B_t = \tilde{A}_t^T, \quad F_{t,s} = -C_t^T N_t^{-1} C_t \tilde{V}_{t-1} W_s^{-1} \tilde{V}_s^T,$$

we obtain

$$Z_{t,s} = Y_{t,s} - \sum_{j=t}^{s-1} Y_{t,j} C_{j+1}^T N_{j+1}^{-1} C_{j+1} \tilde{V}_j W_s^{-1} \tilde{V}_s^T, \quad t = s-1, s-2, \ldots,$$

where

$$Y_{t-1,s} = \tilde{A}_t^T Y_{t-1,s}, \quad Y_{s,s} = I_{l+r}, \quad t = s-1, s-2, \ldots.$$

The use of Lemmas 4.7 and 4.8 gives

$$Y_{t,s} = \tilde{A}_{t+1}^T \tilde{A}_{t+2}^T \ldots \tilde{A}_s^T, \quad \Phi_{t,s} = \tilde{A}_t \tilde{A}_{t+1} \ldots \tilde{A}_{s+1}, \quad \Phi_{s,t}^T = \tilde{A}_{t+1}^T \tilde{A}_{t+2}^T \ldots \tilde{A}_s^T.$$

From this it follows that $Y_{t,s} = \Phi_{s,t}^T$. Thus

$$Z_{t,s} = \Phi_{s,t}^T - \sum_{j=t}^{s-1} \Phi_{j,t}^T C_{j+1}^T N_{j+1}^{-1} C_{j+1} \tilde{V}_j W_s^{-1} \tilde{V}_s^T, \quad t = s-1, s-2, \ldots. \tag{4.121}$$

We have

$$\begin{aligned} G_{t-1,s} &= A_t^T G_{t,s} = (I_{l+r} - K_t C_t)^T G_{t,s} \\ &= ((I_{l+r} - \tilde{S}_t C_t^T R_t^{-1} C_t) - \tilde{V}_s W_t^+ \tilde{V}_s^T C_t^T R_t^{-1} C_t)^T G_{t,s} = (\tilde{A}_t - K_t^d C_t)^T G_{t,s}. \end{aligned} \tag{4.122}$$

From the comparison of the right-hand sides of Eqs. (4.119) and (4.122) it follows that Eq. (4.120) is satisfied if

$$(K_t^d)^T Z_{t,s} = N_t^{-1} C_t \tilde{V}_{t-1} W_s^{-1} \tilde{V}_s^T. \tag{4.123}$$

Using the expressions

$$\tilde{V}_t = \Phi_{t,0} \tilde{V}_0, \quad \Phi_{s,t} \Phi_{t,0} = \Phi_{s,0}$$

gives

$$\begin{aligned} (K_t^d)^T Z_{t,s} &= R_t^{-1} C_t \tilde{V}_t W_t^+ \tilde{V}_t^T \left(\Phi_{s,t}^T - \sum_{j=t}^{s-1} \Phi_{j,t}^T C_{j+1}^T N_{j+1}^{-1} C_{j+1} \tilde{V}_j W_s^{-1} \tilde{V}_s^T \right) \\ &= R_t^{-1} C_t \tilde{V}_t W_t^+ \left(I_{l+r} - \sum_{j=t}^{s-1} \tilde{V}_j^T C_{j+1}^T N_{j+1}^{-1} C_{j+1} \tilde{V}_j W_s^{-1} \right) \overline{V}_s^T. \end{aligned}$$

Iterating Eq. (4.111), we find

$$W_t = \sum_{j=0}^{t-1} \tilde{V}_j^T C_{j+1}^T N_{j+1}^{-1} C_{j+1} \tilde{V}_j.$$

Since

$$\sum_{j=t}^{s-1} \tilde{V}_j^T C_{j+1}^T N_{j+1}^{-1} C_{j+1} \tilde{V}_j = W_s - W_t$$

then the left-hand side of Eq. (4.123) can be transformed to the form

$$\begin{aligned}(K_t^d)^T Z_{t,s} &= R_t^{-1} C_t \tilde{V}_t W_t^+ \left(W_s - \sum_{j=t}^{s-1} \tilde{V}_j^T C_{j+1}^T N_{j+1}^{-1} C_{j+1} \tilde{V}_j \right) W_s^{-1} \tilde{V}_s^T \\ &= R_t^{-1} C_t \tilde{V}_t W_t^+ W_t W_s^{-1} \tilde{V}_s^T.\end{aligned} \tag{4.124}$$

Let us show that

$$R_t^{-1} C_t \tilde{V}_t = N_t^{-1} C_t \tilde{V}_{t-1}. \tag{4.125}$$

Note at first that $\tilde{A}_t = A_{1t}$. Indeed, we have

$$\tilde{A}_t = I_{l+r} - \tilde{S}_t C_t^T R_t^{-1} C_t = \begin{pmatrix} I_l - S_t (C_t^\beta)^T R_t^{-1} C_t^\beta & 0_{l\times r} \\ 0_{r\times l} & I_r \end{pmatrix}, \tag{4.126}$$

where

$$S_t = S_{t-1} - S_{t-1}(C_t^\beta)^T (R_t + C_t^\beta S_{t-1} (C_t^\beta)^T)^{-1} C_t^\beta S_{t-1t}, \quad S_0 = \overline{P}_\beta.$$

Since

$$S_t^{-1} = S_{t-1}^{-1} + (C_t^\beta)^T R_t^{-1} C_t^\beta$$

then using the identity

$$PH^T (HPH^T + R)^{-1} = (P^{-1} + H^T R^{-1} H)^{-1} H R^{-1}$$

with $P = S_{t-1}$, $H = C_t^\beta$, $R = R_t$ gives

$$S_{t-1} (C_t^\beta)^T (R_t + C_t^\beta S_{t-1} (C_t^\beta)^T)^{-1} = S_t (C_t^\beta)^T R_t^{-1}.$$

Taking into account Eq. (4.126), we obtain $\tilde{A}_t = A_{1t}$.

We show now that

$$\tilde{A}_t = I_l - S_t C_t^\beta R_t^{-1} (C_t^\beta)^T = (I_l + S_{t-1} C_t^\beta R_t^{-1} (C_t^\beta)^T)^{-1}. \tag{4.127}$$

Let us transform the right-hand side of this expression using the identity Eq. (2.7). Putting in Eq. (2.7) $B = S_{t-1}$, $C = C_t^{\beta}$, $D = R_t$ we obtain

$$(S_{t-1}^{-1} + C_t^{\beta} R_t^{-1} (C_t^{\beta})^T)^{-1} S_{t-1}^{-1} = (S_{t-1} - S_{t-1}(C_t^{\beta})^T \times \\ \times (R_t + C_t^{\beta} S_{t-1} (C_t^{\beta})^T)^{-1} C_t^{\beta} S_{t-1}) S_{t-1}^{-1} = \\ = I_l - S_{t-1}(C_t^{\beta})^T N_t^{-1} C_t^{\beta} = I_l - S_t (C_t^{\beta})^T R_t^{-1}.$$

Using Eqs. (4.110) and (4.127), we find

$$\tilde{V}_{t-1} = A_{1t}^{-1} \tilde{V}_t = (I_{l+r} + \tilde{S}_{t-1} C_t^T R_t^{-1} C_t) \tilde{V}_t. \quad (4.128)$$

If Eq. (4.125) holds then there should be

$$\tilde{V}_t = \tilde{V}_{t-1} - \tilde{S}_{t-1} C_t^T N_t^{-1} C_t \tilde{V}_{t-1} = \tilde{V}_{t-1} - \tilde{S}_{t-1} C_t^T R_t^{-1} C_t \tilde{V}_t$$

that coincides with Eq. (4.128). Thus we have shown that Eq. (4.125) holds.

Substituting Eq. (4.125) to Eq. (4.124) gives

$$(K_t^d)^T Z_{t,s} = N_t^{-1} C_t \tilde{V}_{t-1} W_t^+ W_t W_s^{-1} \tilde{V}_s^T. \quad (4.129)$$

Let $\tilde{C}_t = (\tilde{V}_0^T C_1^T N_1^{-1/2}, \tilde{V}_1^T C_2^T N_2^{-1/2}, \ldots, \tilde{V}_{t-1}^T C_t^T N_t^{-1/2})$. Assume that the rank of this matrix is equal to $k(t)$ and $l_t(1), l_t(2), \ldots, l_t(k(t))$ are any of its linearly independent columns. The use of the skeletal decomposition yields $\tilde{C}_t = L_t \Gamma_t$, where L_t, Γ_t are $r \times k(t)$, $k(t) \times mt$ matrices of the rank $k(t)$, $\Gamma_t(i)$, $i = 1, 2, \ldots, t$ are $k(t) \times (l + r)$ matrices.

We have

$$W_t = \tilde{C}_t \tilde{C}_t^T = L_t \Gamma_t \Gamma_t^T L_t^T, \quad W_t^+ = (L_t \Gamma_t \Gamma_t^T L_t^T)^+ = (L_t^T)^+ (\Gamma_t \Gamma_t^T)^{-1} (L_t)^+,$$

$$N_t^{-1/2} \tilde{V}_{t-1}^T C_t^T = L_t \tilde{\Gamma}_t,$$

where $L_t^+ = (L_t^T L_t)^{-1} L_t^T$, $\tilde{\Gamma}_t$ is a matrix. Substituting these expressions in Eq. (4.129), we obtain

$$(K_t^d)^T Z_{t,s} = N_t^{-1} C_t \tilde{V}_{t-1} W_t^+ W_t W_s^{-1} \tilde{V}_s^T = R_t^{-1} C_t \tilde{V}_t W_s^{-1} \tilde{V}_s^T. \quad (4.130)$$

This implies Eq. (4.120). Finally, the use of Lemmas 4.7 and 4.8 for $B_t = A_t^T$ gives

$$H_{t,s} = A_t A_{t-1} \ldots A_{s+1}, \quad G_{t,s} = B_{t+1} B_{t+2} \ldots B_s, \quad G_{s,t}^T = B_t^T B_{t-1}^T \ldots B_{s+1}^T.$$

Therefore $H_{t,s} = G_{s,t}^T = Z_{s,t}$ and we have Eq. (4.114). The expression (4.115) follows from Eq. (4.130).

Establish an analogous result for the DTA with forgetting factor described by the system of Eqs. (4.35)–(4.41). We show that in this case we can use Lemma 4.9. Indeed, consider the relations describing the GN method with a linearization in the neighborhood of the last estimate and soft initialization for training with the criterion Eq. (4.4)

$$x_t = x_{t-1} + K_t(y_t - h_t(x_{t-1})),\ \ x_0 = (\overline{\beta}^T, 0_{1\times r})^T,\quad t = 1, 2, \ldots, \tag{4.131}$$

where $h_t(x_{t-1}) = \Phi(z_t, \beta_{t-1})\alpha_{t-1}$,

$$K_t = M_t^{-1} C_t^T = P_t C_t^T, \tag{4.132}$$

$$M_t = \lambda M_{t-1} + C_t^T C_t,\ M_0 = block\ diag(\overline{P}_\beta^{-1}, \overline{P}_\beta^{-1}/\mu),$$

$$P_t = (P_{t-1} - P_{t-1}C_t^T(\lambda I_r + C_t P_{t-1} C_t^T)^{-1} C_t P_{t-1})/\lambda,\ P_0 = block\ diag(\overline{P}_\beta, \overline{P}_\alpha \mu), \tag{4.133}$$

$$C_t^\beta = C_t^\beta(x_{t-1}) = \partial[\Phi(z_t, \beta_{t-1})\alpha_{t-1}]/\partial\beta_{t-1},\ C_t^\alpha = C_t^\alpha(x_{t-1}) = \Phi(z_t, \beta_{t-1}).$$

Transforming K_t by replacing $L_t = \lambda^{-t} M_t$, we get

$$K_t = \lambda^{-t} L_t^{-1} C_t^T, \tag{4.134}$$

where the matrices L_t and L_t^{-1} can be found recursively

$$L_t = L_{t-1} + \lambda^{-t} C_t^T C_t,\ \ L_0 = block\ diag(\overline{P}_\beta^{-1}, \overline{P}_\beta^{-1}/\mu)/\lambda,$$

$$L_t^{-1} = L_{t-1}^{-1} - L_{t-1}^{-1} C_t^T(\lambda^t I_r + C_t L_{t-1}^{-1} C_t^T)^{-1} C_t L_{t-1}^{-1},\ L_o^{-1} = block\ diag(\overline{P}_\beta, \overline{P}_\alpha \mu)\lambda. \tag{4.135}$$

In view of Eqs. (4.134) and (4.135) and Theorem 4.3, we obtain

$$K_t^\beta = (S_t + V_t W_t^+ V_t^T)(C_t^\beta)^T \lambda^{-t} + V_t W_t^+ (C_t^\alpha)^T \lambda^{-t}, \tag{4.136}$$

$$K_t^\alpha = W_t^+ V_t^T (C_t^\beta)^T \lambda^{-t} + W_t^+ (C_t^\alpha)^T \lambda^{-t}, \tag{4.137}$$

$$S_t = S_{t-1} - S_{t-1}(C_t^\beta)^T(\lambda^t I_m + C_t^\beta S_{t-1}(C_t^\beta)^T)^{-1} C_t^\beta S_{t-1},\ \ S_0 = \overline{P}_\beta, \tag{4.138}$$

$$\begin{aligned} V_t = {} & (I_l - S_{t-1}(C_t^\beta)^T(\lambda^t I_m + C_t^\beta S_{t-1}(C_t^\beta)^T)^{-1} C_t^\beta) V_{t-1} \\ & - S_{t-1}(C_t^\beta)^T(\lambda^t I_m + C_t^\beta S_{t-1}(C_t^\beta)^T)^{-1} C_t^\alpha,\ \ V_0 = 0_{l\times r}, \end{aligned} \tag{4.139}$$

$$\begin{aligned} W_t = {} & W_{t-1} + (C_t^\beta V_{t-1} + C_t^\alpha)^T(\lambda^t I_m + C_t^\beta S_{t-1}(C_t^\beta)^T)^{-1}(C_t^\beta V_{t-1} + C_t^\alpha), \\ & W_0 = 0_{r\times r}, \end{aligned} \tag{4.140}$$

where $K_t = ((K_t^\beta)^T, (K_t^\alpha)^T)^T$.

Let us represent these expressions for the DTA in the following more compact and convenient form for use

$$K_t = (\tilde{S}_t + \tilde{V}_t \tilde{W}_t^+ \tilde{V}_t^T) C_t^T \lambda^{-t} = \tilde{S}_t C_t^T \lambda^{-t} + K_t^d, \quad t = 1, 2, \ldots, \tag{4.141}$$

where

$$\tilde{S}_t = \tilde{S}_{t-1} - \tilde{S}_{t-1} C_t^T N_t^{-1} C_t \tilde{S}_{t-1}, \tag{4.142}$$

$$\tilde{V}_t = (I_{l+r} - \tilde{S}_{t-1} C_t^T N_t^{-1} C_t) \tilde{V}_{t-1} = A_{1t} \tilde{V}_{t-1}, \tag{4.143}$$

$$\tilde{W}_t = \tilde{W}_{t-1} + \tilde{V}_{t-1}^T C_t^T N_t^{-1} C_t \tilde{V}_{t-1}, \tag{4.144}$$

$$N_t = \lambda^t I_m + C_t \tilde{S}_{t-1} C_t^T. \tag{4.145}$$

Assume $H_{t,s}$ is the matrix function determined by the system

$$H_{t,s} = (I_{l+r} - K_t C_t) H_{t-1,s} = A_t H_{t-1,s}, \quad t = s+1,\ s+2, \ldots, \quad H_{s,s} = I_{l+r}, \tag{4.146}$$

where

$$K_t = M_t^{-1} C_t^T = (\tilde{S}_t + \tilde{V}_t \tilde{W}_t^+ \tilde{V}_t^T) C_t^T \lambda^{-t} = \tilde{S}_t C_t^T \lambda^{-t} + K_t^d. \tag{4.147}$$

It follows from Lemma 4.9 that

$$H_{t,s} = \Phi_{t,s} - \tilde{V}_t L_t^{-1} \sum_{i=1}^{t-1} \tilde{V}_i^T C_{i+1}^T N_{i+1}^{-1} C_{i+1} \Phi_{i,s}, \tag{4.148}$$

$$H_{t,s} K_s^d = \tilde{V}_t L_t^{-1} \tilde{V}_s^T C_s^T \lambda^{-s}, \tag{4.149}$$

where $\Phi_{t,s}$ is the solution of the system

$$\Phi_{t,s} = (I_{l+r} - \lambda^{-t} \tilde{S}_t C_t^T C_t) \Phi_{t-1,s} = \tilde{A}_t \Phi_{t-1,s}, \quad \Phi_{s,s} = I_{l+r}. \tag{4.150}$$

We turn now to the analysis of the convergence of the DTA.

Theorem 4.5.

Let the following conditions hold:

1. $P(\|\xi_t\| < \infty) = 1$ for $t \in T$, where $T = \{1, 2, \ldots, N\}$ is a bounded set.
2. $\Phi(z_t, \beta) \in C_{(z,\beta)}^{(0,2)}(Z \times A_\beta)$, $z_t \in Z \subset R^n$ for $t \in T$, $\beta \in A_\beta = \{\beta : \|\beta - \beta^*\| \le \delta_\beta\}$, $\delta_\beta > 0$ is a number, Z is an arbitrary compact set.
3. There is $tr = \min\{t : W_t > 0, t \in T\}$ and the rank of the matrix W_t is a constant for each fixed $t = 1, 2, \ldots, tr - 1$.

Then the solution $z_t = x^*$ of the system Eq. (4.99) is finitely stable under the action of the disturbance ξ_t.

Moreover, for every $tf \geq tr$, $tf \in T$ there are $\rho_0 > 0$, $\chi > 0$, $\varsigma > 0$ such that for $t \in Tf = \{tf, tf+1, \ldots, N\}$ and $\rho < \rho_0$ will be valid uniform in $\alpha^* \in A_d$ the bound

$$||e_t|| = ||x_t - x^*|| \leq (\chi||e_0^\beta|| + \varsigma \max_{t \in \{1,2,\ldots,tf\}} ||\xi_t||)(1+\nu(\rho))^{t-1}, \quad (4.151)$$

where $\rho = \max\{||e_0^\beta||, \ ||S_0||, \ \max_{t \in \{1,2,\ldots,tf\}} ||\xi_t||\}$, $\nu(\rho) = O(\rho)$ as $\rho \to 0$.

Proof

Consider at first the DTA, taking into account the matrix R_t. Transform $h_t(x^*)$in Eq. (4.99) using the formula [70]

$$f(x+y) = f(x) + (\nabla f(x), y) + y^T \nabla^2 f(x + \theta y) y/2,$$

where $\nabla f(x)$ and $\nabla^2 f(x)$ are the gradient and Hessian matrix of $f(x)$, respectively, $0 \leq \theta \leq 1$.

Let $f(x) = h_{ti}(x)$, $x = x_{t-1}$, $y = x^* - x_{t-1} = -e_{t-1}$. Then

$$h_{it}(x^*) = h_{it}(x_{t-1}) - C_{it}(x_{t-1}) e_{t-1} + \varphi_{it}(x_{t-1}, x^*), \quad i = 1, 2, \ldots, m, \quad (4.152)$$

where

$$C_t = C_t(x_{t-1}) = (C_{1t}^T, C_{2t}^T, \ldots, C_{mt}^T)^T,$$

$$\varphi_{it} = \varphi_{it}(x_{t-1}, x^*) = e_{t-1}^T \nabla^2 h_{it}(x_{t-1} - \theta e_{t-1}) e_{t-1}/2, \quad i = 1, 2, \ldots, m.$$

Using Eq. (4.152) gives

$$e_t = (I_{l+r} - K_t C_t) e_{t-1} + K_t(\varphi_t + \xi_t), \quad e_0 = -x^*, \quad t \in T, \quad (4.153)$$

where $\varphi_t = (\varphi_{1t}, \varphi_{2t}, \ldots, \varphi_{mt})^T$. It follows from this

$$e_t = H_{t,0} e_0 + \sum_{s=1}^{t} H_{t,s} K_s(\varphi_s + \xi_s), \quad (4.154)$$

where $H_{t,s}$ is a matrix determined by the system

$$H_{t,s} = (I_{l+r} - K_t C_t) H_{t-1,s}, \quad t > s, \quad H_{t,t} = I_{l+r}. \quad (4.155)$$

Using condition 2 and Lemma 4.9 gives for $t \geq tr$, $t \in T$

$$\lim_{\rho \to 0} C_t = \lim_{\rho \to 0} C_t(x_{t-1}) = (\partial \Phi(z_t, \beta^*)/\partial \beta^* \alpha^*, \Phi(z_t, \beta^*)), \quad (4.156)$$

$$\lim_{\rho\to 0} S_t = 0, \tag{4.157}$$

$$\lim_{\rho\to 0} H_{t,s} = \lim_{\rho\to 0}\left[\Phi_{t,s} - \tilde{V}_t W_t^{-1}\sum_{i=s}^{t-1}\tilde{V}_i^T C_{i+1}^T N_{i+1}^{-1} C_{i+1}\Phi_{i,s}\right]$$
$$= \begin{pmatrix} I_l & 0_{l\times r} \\ 0_{r\times l} & I_r - (W_t^*)^{-1}\sum_{j=s}^{t-1}\Phi^T(z_{j+1},\beta^*)R_{j+1}^{-1}\Phi(z_{j+1},\beta^*) \end{pmatrix}, \tag{4.158}$$

$$\lim_{\rho\to 0} H_{t,s}K_s^d = \lim_{\rho\to 0}\tilde{V}_t W_t^{-1}\tilde{V}_s^T C_s^T R_s^{-1} = \begin{pmatrix} 0_{1\times m} \\ (W_t^*)^{-1}\Phi^T(z_s,\beta^*)R_s^{-1} \end{pmatrix}, \tag{4.159}$$

$$\lim_{\rho\to 0} H_{t,s}\tilde{S}_s C_s^T R_s^{-1} = 0, \tag{4.160}$$

where $W_t^* = \sum_{j=0}^{t-1}\Phi^T(z_{j+1},\beta^*)R_{j+1}^{-1}\Phi(z_{j+1},\beta^*)$.

It follows from Lemma 4.9 that for $t \geq tr$

$$H_{t,0}(0_{1\times m},(e_0^\alpha)^T)^T = H_{t,0}\tilde{V}_0 e_0^\alpha = 0$$

and thus

$$H_{t,0}e_0 = H_{t,0}((e_0^\beta)^T, 0_{1\times r})^T. \tag{4.161}$$

Therefore, for any $tf \geq tr$, $tf \in T$ there exist positive constants ρ_0, η_1, η_2 such that for $t \in Tf$, $\rho < \rho_0$ and all $\alpha^* \in A_d$ the following estimates uniform in $\alpha^* \in A_d$ are valid

$$||H_{t,0}e_0|| \leq \eta_1||e_0^\beta||,\ \ ||H_{t,s}K_s|| \leq \eta_2,\ \ ||H_{t,s}K_s\xi_s|| \leq \eta_2 \max_{s\in\{1,2,\ldots,tf\}}||\xi_s||. \tag{4.162}$$

Using Eqs. (4.154) and (4.162) gives

$$||e_t|| \leq \eta_1||e_0^\beta| + \eta_2\sum_{s=1}^{t}||\varphi_s|| + \eta_2 N \max_{t\in\{1,2,\ldots,tf\}}||\xi_t||,\quad t\in Tf. \tag{4.163}$$

Let us now get the bounds for φ_{it}, $i=1,2,\ldots,m$, $t\in\{1,2,\ldots,tf\}$. We have

$$\varphi_{it}=\left((e_{t-1}^{\beta})^T,(e_{t-1}^{\alpha})^T\right)\begin{pmatrix}H_{it}^1 & H_{it}^2\\ (H_{it}^2)^T & 0_{r\times r}\end{pmatrix}\left((e_{t-1}^{\beta})^T,(e_{t-1}^{\alpha})^T\right)^T/2$$
$$=(e_{t-1}^{\beta})^T H_{it}^1 e_{t-1}^{\beta}/2+(e_{t-1}^{\beta})^T(H_{it}^2)^T e_{t-1}^{\alpha},\quad i=1,2,\ldots,m,$$

where

$$H_{it}^1=\partial^2\Phi_i(z_t,\beta_{t-1}-\theta e_{t-1}^{\beta})/\partial\beta_{t-1}^2(\alpha_{t-1}-\theta e_{t-1}^{\alpha})$$
$$=\partial^2\Phi_i(z_t,\beta^*-(1-\theta)e_{t-1}^{\beta})/\partial\beta_{t-1}^2(\alpha^*-(1-\theta)e_{t-1}^{\alpha})\in R^{l\times l},$$
$$H_{it}^2=\partial\Phi_i(z_t,\beta_{t-1}-\theta e_{t-1}^{\beta})/\partial\beta_{t-1}=\partial\Phi_i(z_t,\beta^*-(1-\theta)e_{t-1}^{\beta})/\partial\beta_{t-1}\in R^{l\times r}.$$

Or in a more compact form

$$\varphi_{it}=\zeta_{it}(x_{t-1},x^*)e_{t-1}^{\beta}+\psi_{it}(x_{t-1},x^*)e_{t-1}^{\alpha}$$
$$=(\zeta_{it}(x_{t-1},x^*),\psi_{it}(x_{t-1},x^*))e_{t-1},\quad i=1,2,\ldots,m,$$

where

$$\zeta_{it}(x_{t-1},x^*)=(e_{t-1}^{\beta})^T\partial^2\Phi_i(z_t,\beta^*-(1-\theta)e_{t-1}^{\beta})/\partial\beta_{t-1}^2\alpha^*/2,\tag{4.164}$$

$$\psi_{it}(x_{t-1},x^*)=(e_{t-1}^{\beta})^T[\partial\Phi_i(z_t,\beta^*-(1-\theta)e_{t-1}^{\beta})/\partial\beta_{t-1}-\partial^2\Phi_i(z_t,\beta^*-(1-\theta)e_{t-1}^{\beta})/\partial\beta_{t-1}^2(1-\theta)/2].\tag{4.165}$$

It follows from conditions 2 and 3 the existence of the limits for $t\in T$

$$\lim_{\rho\to 0}\partial^2\Phi_i(z_t,\beta_{t-1}-\theta e_{t-1}^{\beta})/\partial\beta_{t-1}^2\alpha_{t-1}=\partial^2\Phi_i(z_t,\beta)/\partial\beta^2|_{\beta=\beta^*}\alpha^*,\tag{4.166}$$

$$\lim_{\rho\to 0}\partial\,\Phi_i(z_t,\beta_{t-1}-\theta e_{t-1}^{\beta})/\partial\beta_{t-1}=\partial\Phi_i(z_t,\beta^*)/\partial\beta^*,\tag{4.167}$$

$$\lim_{\rho\to 0}e_t^{\beta}=0.\tag{4.168}$$

Therefore, for any $\alpha^*\in A_d$ there exists $\rho_0>0$ such that uniformly in $\alpha^*\in A_d$ and $t\in\{1,2,\ldots,tf\}$

$$||\varphi_t||\le\upsilon(\rho)||e_{t-1}||,\quad t\in T,\tag{4.169}$$

where $\upsilon(\rho)=O(\rho)$ as $\rho\to 0$. Thus

$$||e_t||\le\eta_1||e_0^{\beta}|+\eta_2\upsilon(\rho)\sum_{s=1}^{t}||e_{s-1}||+\eta_2 N\max_{t\in\{1,2,\ldots,tf\}}||\xi_t||,\quad t\in Tf.\tag{4.170}$$

We use Lemma Bellman to assess $||e_t||$ [71]: if

$$z_t \le z_0 + q\sum_{s=1}^{t-1} z_s,$$

where $z_t > 0$, $q > 0$ then

$$z_t \le z_0(1+q)^{t-1}, \quad t = 1, 2, \ldots, N.$$

Putting $z_t = ||e_t||$, $z_0 = \eta_1 ||e_0^{\beta}| + \eta_2 N \max_{t\in\{1,2,\ldots,tf\}} ||\xi_t||$, $q = \eta_2 \upsilon(\rho)$, we obtain Eq. (4.151).

The proof for the DTA with the forgetting factor is carried out in the same way. So, in this case, instead of Lemma 4.9 the relations (4.148) and (4.149) are used to establish the existence of the limits

$$\lim_{\rho\to 0} H_{t,s} = \lim_{\rho\to 0}\left[\Phi_{t,s} - \tilde{V}_t L_t^{-1} \sum_{i=s}^{t-1} \tilde{V}_i^T C_{i+1}^T N_{i+1}^{-1} C_{i+1} \Phi_{i,s}\right]$$

$$= \begin{pmatrix} I_l & 0_{l\times r} \\ 0_{r\times l} & I_r - (W_t^*)^{-1} \sum_{j=s}^{t-1} \lambda^{-j-1} \Phi^T(z_{j+1},\beta^*)\Phi(z_{j+1},\beta^*) \end{pmatrix},$$

$$\lim_{\rho\to 0} H_{t,s} K_s^d = \lim_{\rho\to 0} \tilde{V}_t L_t^{-1} \tilde{V}_s^T C_s^T \lambda^{-s} = \begin{pmatrix} 0_{1\times m} \\ (W_t^*)^{-1} \varPhi^T(z_s,\beta^*)\lambda^{-s} \end{pmatrix},$$

$$\lim_{\rho\to 0} H_{t,s} \tilde{S}_s C_s^T = 0,$$

and the bounds Eq. (4.162), where $W_t^* = \sum_{j=0}^{t-1} \lambda^{-j-1} \Phi^T(z_{j+1},\beta^*)$ $\Phi(z_{j+1},\beta^*)$.

Consider the two-stage DTA described by the expressions (4.35) and (4.68)–(4.70). Let us show that Theorem 4.5 is saved in this case. It follows from Eqs. (4.97), (4.98), and (4.152) the following equations

$$e_t^{\beta} = (I_l - K_t^{\beta} C_t^{\beta}) e_{t-1}^{\beta} + K_t^{\beta}(\varphi_t + \xi_t), \ \ e_0^{\beta} = \overline{\beta} - \beta^*,$$

$$e_t^{\alpha} = (I_l - K_t^{\alpha} C_t^{\beta}) e_{t-1}^{\beta} + K_t^{\alpha}(\varphi_t + \xi_t), \ \ e_0^{\alpha} = -\alpha^*$$

iterating which, we obtain

$$e_t^{\beta} = H_{t,0}^{\beta} e_0^{\beta} + \sum_{s=1}^{t} H_{t,s}^{\beta} K_s^{\beta} (\varphi_s + \xi_s),$$

$$e_t^\alpha = H_{t,0}^\alpha e_0^\alpha + \sum_{s=1}^{t} H_{t,s}^\alpha K_s^\alpha(\varphi_s + \xi_s),$$

where

$$H_{t,s}^\beta = (I_l - K_t^\beta C_t^\beta)H_{t,s-1}^\beta \quad t > s, \; H_{t,t}^\beta = I_l,$$
$$H_{t,s}^\alpha = (I_r - K_t^\alpha C_t^\alpha)H_{t,s-1}^\beta \quad t > s, \; H_{t,t}^\alpha = I_r.$$

Using the expressions

$$H_{t,s}^\beta = \lambda^{t-s} S_t S_s^{-1}, \; H_{t,s}^\alpha K_t^\alpha = \lambda^{t-s} W_t^+ (C_t^\alpha)^T, \; H_{t,0}^\alpha = l\overline{P}(I_r - W_t W_t^+)$$

and the bounds Eqs. (4.164)–(4.169), we find Eq. (4.170). This implies Eq. (4.151).

4.4.2 Infinite Training Set

Let us obtain conditions for the DTA convergence as $t \to \infty$. The convergence is interpreted as stability of the solution $z_t = x^*$ of the system Eq. (4.99).

Definition 2

The solution of the system Eq. (4.99) $z_t = x^*$ is asymptotically stable if there exists such $\delta(A_d) > 0$ that conditions $||\beta_0 - \beta^*|| < \delta(A_d)$, $\alpha_0 = 0$ imply $\lim_{t\to\infty} ||x_t - x^*|| = 0$ for all $\alpha \in A_d = \{\alpha: ||\alpha - \alpha^*|| \le d\} \subset R^r$ and any given $d > 0$.

Definition 3

The solution of the system Eq. (4.99) $z_t = x^*$ is stable under the action of the disturbance ξ_t if for any $\varepsilon > 0$ there exist such $\delta(A_d) > 0$, $\delta_2(A_d) > 0$ that the conditions $||\beta_0 - \beta^*|| < \delta_1(A_d)$, $\alpha_0 = 0$, $||\xi_t|| < \delta_2(A_d)$ imply $||x_t - x^*|| < \varepsilon$ for $t \ge tr$, all $\alpha \in A_d = \{\alpha: ||\alpha - \alpha^*|| \le d\} \subset R^r$ and any given $d > 0$.

Theorem 4.6.

Suppose that:

1. The conditions of Theorem 4.5 are satisfied.
2. There are positive constants k_1, k_2 such that

$$||\Phi(z_t, \beta)|| \le k_1, \quad ||\partial\Phi(z_t, \beta)/\partial\beta|| \le k_2,$$

for $z_t \in Z \subset R^n$, $\beta \in A_\beta$, $t \in T$, where $A_\beta = \{\beta: ||\beta - \beta^*|| \leq \delta_\beta\}$, $\delta_\beta > 0$ is a number, Z is arbitrary compact set, $T = \{1, 2, \ldots, \}$.

3. For some integer s and all j there exist numbers p_1 and p_2 such that

$$0 < p_1 I_{r+l} \leq \sum_{i=j}^{j+s} C_i^T C_i \leq p_2 I_{r+l} < \infty \tag{4.171}$$

for any $(\beta_t, \alpha_t) \in A_\beta \times A_\alpha$, where $A_\alpha = \{\alpha: ||\alpha - \alpha^*|| \leq \delta_\alpha\} \subset R^r$.

Then: 1. The solution of the system Eq. (4.99) $z_t = x^*$ with K_t described by the expressions (4.36)–(4.41) is asymptotically stable and stable under the action of the disturbance ξ_t.

2. There are positive constants σ, q_1, q_2, τ, ρ_0 such that for $t \in T$, $\rho < \rho_0$ uniformly in $\alpha \in A_d$

$$||e_t||^2 \leq (1 - \lambda + \sigma)^t q_2/q_1 ||e_{tf}||^2 + \tau/q_1 \sup_{t \geq tf} ||\xi_t||^2/(\lambda - \sigma), \tag{4.172}$$

where $tf \geq tr = \min\{t: W_t > 0, t \in T\}$, $\rho = \max\{||e_0^\beta||, \ ||S_0||, \max_{t \in \{1,2,\ldots,tf\}} ||\xi_t||\}$, $||e_{tf}||$ satisfies Eq. (4.151).

Proof

It follows from Theorem 4.5 that for any $tf \geq tr$, $N > tf$, $\delta_\beta > 0$, $\delta_\alpha > 0$ there exists such $\rho_0 > 0$ that if $\rho < \rho_0$ and $t \in T = \{tf, tf + 1, \ldots, N\}$ then $(\beta_t, \alpha_t) \in A_\beta \times A_\alpha$.

Consider the Lyapunov function in the area $A_\beta \times A_\alpha$

$$V_t = e_t^T P_t^{-1} e_t,$$

where P_t^{-1} is determined by the system

$$P_t^{-1} = \lambda P_{t-1}^{-1} + (C_t)^T C_t$$

for $t \geq tf \in T$ with the initial condition $P_{tf}^{-1} > 0$.

Transforming V_t with help of the relation for the error Eq. (4.153), we obtain

$$\begin{aligned} V_t = e_{t-1}^T P_t^{-1} e_t = e_t^T (I_{l+r} - K_t C_t)^T P_t^{-1} (I_{l+r} - K_t C_t) e_{t-1} \\ + 2e_{t-1}^T (I_{l+r} - K_t C_t)^T P_t^{-1} K_t \varphi_t - 2e_{t-1}^T (I_{l+r} - K_t C_t)^T P_t^{-1} K_t \xi_t \\ + 2\xi_t^T K_t^T P_t^{-1} K_t \varphi_t + \xi_t^T K_t^T P_t^{-1} K_t \xi_t + \varphi_t^T K_t^T P_t^{-1} K_t \varphi_t. \end{aligned} \tag{4.173}$$

Now we find bounds for terms in Eq. (4.173). As $K_t = P_t C_t$ and P_t satisfies the system

$$P_t = P_{t-1}/\lambda - P_{t-1} C_t^T (\lambda I_m + C_t P_{t-1} C_t^T)^{-1} C_t P_{t-1}/\lambda$$

for $t \geq tf \in T$ with the initial condition $P_{tf}^{-1} > 0$ then

$$\begin{aligned}
&e_{t-1}^T (I_{l+r} - K_t C_t)^T P_t^{-1} (I_{l+r} - K_t C_t) e_{t-1} \\
&\quad = e_{t-1}^T (I_{l+r} - P_t C_t^T C_t)^T P_t^{-1} (I_{l+r} - P_t C_t^T C_t) e_{t-1} \\
&\quad = e_{t-1}^T (P_t^{-1} - C_t^T C_t)^T (I_{l+r} - P_t C_t^T C_t) e_{t-1} \\
&\quad = \lambda e_{t-1}^T P_{t-1}^{-1} (I_{l+r} - P_t C_t^T C_t) e_{t-1} = \lambda^2 e_{t-1}^T P_{t-1}^{-1} P_t P_{t-1}^{-1} e_{t-1} \\
&\quad = \lambda e_{t-1}^T [P_{t-1}^{-1} - C_t^T (\lambda I_m + C_t P_{t-1} C_t^T)^{-1} C_t] e_{t-1} \leq \lambda e_{t-1}^T P_{t-1}^{-1} e_{t-1}.
\end{aligned} \tag{4.174}$$

From Eqs. (4.164) to (4.168) it follows that for any $\nu > 0$ there exists such $\delta > 0$ that $||\varphi_t|| \leq \nu| \, ||e_{t-1}||$ until $||e_{t-1}|| \leq \delta$ and from Ref. [72] existence of $q_1 > 0$ and $q_2 > 0$ such that

$$q_1 I_{r+l} \leq P_t^{-1} \leq q_2 I_{r+l}.$$

Therefore, there are $\eta_1 > 0$, $\eta_2 > 0$, $\eta_3 > 0$ such that

$$\varphi_t^T K_t^T P_t^{-1} K_t \varphi_t \leq \eta_1 \nu^2 e_{t-1}^T e_{t-1}, \tag{4.175}$$

$$\xi_t^T K_t^T P_t^{-1} K_t \xi_t \leq \eta_2 \xi_t^T \xi_t, \tag{4.176}$$

$$2 e_{t-1}^T (I_{l+r} - K_t C_t)^T P_t^{-1} K_t \varphi_t \leq \eta_3 \nu e_{t-1}^T e_{t-1}. \tag{4.177}$$

Using Young's inequality [73]

$$2 x^T y \leq x^T X x + y^T X^{-1} y,$$

for $x = \xi_t$, $y = P_t^{-1} (I_{l+r} - K_t C_t) e_{t-1}$, $X = I_{l+r} / \vartheta$, we obtain

$$\begin{aligned}
&2 \xi_t^T P_t^{-1} (I_{l+r} - K_t C_t) e_{t-1} \leq \xi_t^T \xi_t / \vartheta + \vartheta e_{t-1}^T P_t^{-2} (I_{l+r} - K_t C_t)^2 e_{t-1} \leq \\
&\quad \leq \xi_t^T \xi_t / \vartheta + \eta_4 \vartheta e_{t-1}^T e_{t-1}, \quad \eta_4 > 0,
\end{aligned}$$

where $\vartheta > 0$ is a small number.

Putting in Young's inequality $x = \xi_t$, $y = K_t^T P_t^{-1} K_t \varphi_t$, $X = I_{l+r}$ and using Eq. (4.175), we obtain

$$\begin{aligned}
&2 \xi_t^T K_t^T P_t^{-1} K_t \varphi_t \leq \xi_t^T \xi_t + \varphi_{t-1}^T (K_t^T P_t^{-1} K_t)^2 \varphi_{t-1}. \\
&\quad \leq \xi_t^T \xi_t + \eta_5 \nu \, e_{t-1}^T e_{t-1}, \quad \eta_5 > 0.
\end{aligned} \tag{4.178}$$

Using these bounds gives

$$\begin{aligned}
V_t &\leq \lambda e_{t-1}^T P_{t-1}^{-1} e_{t-1} + (\eta_1 \nu^2 + \eta_3 \nu + \eta_4 \vartheta + \eta_{4\backslash 5} \nu) e_{t-1}^T e_{t-1} + (1 + \eta_2 + 1/\vartheta) \xi_t^T \xi_t \\
&\leq (\lambda - \sigma) V_{t-1} + \eta_6 \sup_{t \geq tf} ||\xi_t||^2
\end{aligned}$$

as long as $||e_t|| \leq \delta$ and $(\beta_t, \alpha_t) \in A_\beta \times A_\delta$, where $\sigma = (\eta_1 \nu^2 + \eta_3 \nu + \eta_4 \vartheta + \eta_{4\backslash 5} \nu)/q_1$, $\eta_6 = 1 + \eta_2 + 1/\vartheta$.

Choosing ϑ, ν so that $\lambda > \sigma$, we obtain

$$q_1 ||e_t||^2 \leq V_t \leq (1-\lambda+\sigma)^t V_{tf} + \eta_6 \, \sup_{t \geq tf} ||\xi_t||^2 \sum_{i=0}^{\infty} (1-\lambda+\sigma)^i$$
$$\leq (1-\lambda+\sigma)^t q_2 ||e_{tf}||^2 + \eta_6 \, \sup_{t \geq tf} ||\xi_t||^2 / (\lambda + \sigma).$$

It follows from this

$$||e_t||^2 \leq (1-\lambda+\sigma)^t q_2/q_1 ||e_{tf}||^2 + \eta_6/q_1 \, \sup_{t \geq tf} ||\xi_t||^2 / (\lambda + \sigma). \quad (4.179)$$

We choose $||e_{tf}||$ and restrictions on $\sup_{t \geq tf} ||\xi_t||^2$ to satisfy the condition

$$(1-\lambda+\sigma)^{tf} q_2/q_1 ||e_{tf}||^2 + \eta_6/q_1 \, \sup_{t \geq tf} ||\xi_t||^2 / (\lambda + \sigma) \leq \delta^2.$$

From Theorem 4.5 it follows that it is always possible if ρ is sufficiently small. With this choice the inequality Eq. (4.179) will be carried out for all $t > tf$.

From Eq. (4.179) follows the stability of the solution $x_t = x^*$ of the system Eq. (4.99) under the action of the disturbance ξ_t and if $\xi_t = 0$ then its asymptotic stability.

Theorem 4.7.

Suppose that:

1. The conditions of Theorem 4.5 are satisfied.
2. There are positive constants k_1, k_2 such that

$$||\Phi(z_t, \beta)|| \leq k_1, \quad ||\partial\Phi(z_t, \beta)/\partial\beta|| \leq k_2,$$

   for $z_t \in Z \subset R^n$, $\beta \in A_\beta$, $t \in T$, where $A_\beta = \{\beta : ||\beta - \beta^*|| \leq \delta_\beta\}$, $\delta_\beta > 0$ is a number, Z is a compact set, $T = \{1, 2, \ldots, \}$.
3. $\lim_{t \to \infty} \sum_{i=tf+1}^{t} C_i^T R_i^{-1} C_i / t = P_\infty^{-1}$ for $(\beta_t, \alpha_t) \in A_\beta \times A_\alpha$, where $A_\alpha = \{\alpha : ||\alpha - \alpha^*|| \leq \delta_\alpha\}$, $\delta_\alpha > 0$ is a number, $tf \geq tr$.
4. There is $k_4 > 0$ such that $R_t \geq k_4 I_m$ for $t \in T$. Then: The solution of system Eq. (4.99) $x_t = x^*$ with the matrix K_t described by the expressions (4.59)–(4.64) is asymptotically stable and stable under the action of the disturbance ξ_t.

Proof

Let us at first show that

$$P_t = \tilde{S}_t + \tilde{V}_t W_t^{-1} \tilde{V}_t^T > 0 \quad (4.180)$$

for $t \geq tr \in T$, where

$$\tilde{S}_t = \begin{pmatrix} S_t & 0_{l\times r} \\ 0_{r\times l} & 0_{r\times r} \end{pmatrix}, \quad \tilde{V}_t = \begin{pmatrix} V_t \\ I_r \end{pmatrix}.$$

We represent Eq. (4.180) in the equivalent form

$$P_t = \begin{pmatrix} S_t + V_t W_t^{-1} V_t^T & V_t W_t^{-1} \\ W_t^{-1} V_t^T & W_t^{-1} \end{pmatrix} > 0.$$

As $S_t + V_t W_t^{-1} V_t^T > 0$ then [73]

$$|P_t| = |S_t + V_t W_t^{-1} V_t^T| \times |W_t^{-1} - W_t^{-1} V_t^T (S_t + V_t W_t^{-1} V_t^T)^{-1} V_t W_t^{-1}|.$$

Transforming the second factor in this expression using the identity Eq. (2.7) for $B = W_t^{-1}$, $C = V_t^T$, $D = S_t$, we obtain

$$|P_t| = |S_t + V_t W_t^{-1} V_t^T| \times |(W_t + V_t^T S_t^{-1} V_t)^{-1}|.$$

But since for any square matrix $|A^{-1}| = 1/|A|$ then

$$|P_t| = |S_t + V_t W_t^{-1} V_t^T| / |W_t + V_t^T S_t^{-1} V_t| > 0.$$

From this it follows that P_t satisfies the system

$$P_t = P_{t-1} - P_{t-1} C_t^T (R_t + C_t P_{t-1} C_t^T)^{-1} C_t P_{t-1} \tag{4.181}$$

for $t \geq tf \in T$ with the initial condition $P_{tr} > 0$.

We have for the estimation error

$$e_t = H_{t,tf} e_{tf} + \sum_{s=tf+1}^{t} H_{t,s} K_s (\varphi_s + \xi_s),$$

where $H_{t,s}$ is the system solution Eq. (4.155), φ_t is described by Eq. (4.152), $tf > tr$. Since

$$H_{t,s} = P_t P_s^{-1}, \quad K_t = P_t C_t^T R_t^{-1}$$

for $t \geq s \geq tr$ then

$$e_t = P_t P_{tf}^{-1} e_{tf} + P_t \sum_{s=tf+1}^{t} C_s^T R_s^{-1} (\varphi_s + \xi_s). \tag{4.182}$$

It follows from Theorem 4.5 that for any $tf \geq tr$, $N > tf$, $\delta_\alpha > 0$ there exists such $\rho_0 > 0$ that if $\rho < \rho_0$ and $t \in T = \{tf, tf+1, \ldots, N\}$ then $(\beta_t, \alpha_t) \in A_\beta \times A_\alpha$, where

$$\rho = \max\{\|e_0^\beta\|, \|S_0\|, \sup_{t\in T}\|\xi_t\|\}.$$

From Eqs. (4.164) to (4.168) it follows that for any $\nu > 0$ there exists $\delta > 0$ such that

$$||\varphi_t|| \leq \nu| \ ||e_t||$$

as long as $||e_t|| \leq \delta$ and at the same time

$$||e_t|| \leq ||P_t|| \ ||P_{tf}^{-1}|| \ ||e_{tf}|| + \eta_1||P_t||\left(\nu \sum_{s=tf+1}^{t} ||e_{s-1}|| + \sup_{t\in T}||\xi_t||\right), \tag{4.183}$$

where $\eta_1 = \sup_{t\in T}||C_t^T R_t^{-1}||$ for $(\beta_t, \alpha_t) \in A_\beta \times A_\alpha$.

Using Eq. (4.181) and condition 3 provides $(\beta_t, \alpha_t) \in A_\beta \times A_\alpha$

$$\lim_{t\to\infty} P_t^{-1}/t = \lim_{t\to\infty}\left(P_{tr}^{-1} + \sum_{i=tr+1}^{t} C_i^T R_i^{-1} C_i\right)/t = P_\infty^{-1}. \tag{4.184}$$

Thus for sufficiently large $tf \geq tr$ exists $\eta_2 > 0$ such that

$$||P_t|| = ||P_\infty + \Delta P_t||/t \leq \eta_2/t. \tag{4.185}$$

for any $(\beta_t, \alpha_t) \in A_\beta \times A_\alpha$, where $\Delta P_t \to 0$ as $t \to \infty$. Substituting Eq. (4.185) in Eq. (4.183) yields

$$\begin{aligned} ||e_t|| &\leq \eta_2||P_{tf}^{-1}|| \ ||e_{tf}||/t + \eta_1\eta_2\left(\nu/t \sum_{s=tf+1}^{t} ||e_{s-1}|| + \sup_{t\geq tf+1}||\xi_t||\right) \\ &= \eta_3/t + \eta_4/t \sum_{s=tf+1}^{t} ||e_{s-1}|| + \eta_5, \end{aligned} \tag{4.186}$$

where $\eta_3 = \eta_2 \ ||P_{tf}^{-1}|| \ ||e_{tf}||$, $\eta_4 = \eta_1\eta_2\nu$, $\eta_5 = \eta_1\eta_2 \sup_{t\in T}||\xi_t||$.

From the comparison principle [74] it follows that $||e_t|| \leq u_t$, where u_t is the solution of the difference equation

$$u_t = \eta_3/t + \eta_4/t \sum_{s=tf+1}^{t} u_{s-1} + \eta_5 \tag{4.187}$$

with the initial condition $u_{tf} = \eta_3/tf$.

We have

$$u_t = \eta_3/t + \eta_4/t \sum_{s=tf+1}^{t-1} u_{s-1} + \eta_4 u_{t-1}/t + \eta_5, \tag{4.188}$$

$$u_{t-1} = \eta_3/(t-1) + \eta_4/(t-1) \sum_{s=tf+1}^{t-1} u_{s-1} + \eta_5. \tag{4.189}$$

From Eq. (4.189) we find

$$\sum_{s=tf+1}^{t-1} u_{s-1} = (t-1)/\eta_4(u_{t-1} - \eta_3/(t-1) - \eta_5).$$

Substituting this expression in Eq. (4.188) gives

$$\begin{aligned} u_t &= \eta_3/t + (t-1)/t(u_{t-1} - \eta_3\ /(t-1) - \eta_5) + \eta_4 u_s/t + \eta_5 \\ &= [1 - (1-\eta_4\ )/t]u_{t-1} + \eta_5/t. \end{aligned} \tag{4.190}$$

It follows from this that

$$u_t = h_{t,tf}\eta_3/tf + \eta_5 \sum_{s=tf+1}^{t} h_{t,s}/s, \tag{4.191}$$

where $h_{t,s}$ is the transition function determined by the equation

$$h_{t,s} = [1 - (1-\eta_4\ )/t]h_{t,s-1},\quad h_{s,s} = 1.$$

We have

$$h_{t,s} = \prod_{j=s+1}^{t} [1 - (1-\eta_4\ )/j] \le (1+q_s)(s/t)^{1-\eta_4}. \tag{4.192}$$

for $\eta_4 < 1$ [75], where $q_s \to 0$ as $s \to \infty$.

Substituting Eq. (4.192) in Eq. (4.191) yields

$$u_t \le \eta_3(1+q_{tf})tf^{-\eta_4}(1/t)^{1-\eta_4} + \eta_5 \sum_{s=tf+1}^{t} (1+q_s)(s/t)^{1-\eta_4}/s.$$

We obtain for the second term in this expression

$$\begin{aligned} &\eta_5 \sum_{s=tf+1}^{t} (1+q_s)(s/t)^{1-\eta_4}/s \le \eta_5\ \sup_{s \ge tf+1}(1+q_s)\frac{\sum_{s=tf+1}^{t} s^{-\eta_4}}{t^{1-\eta_4}} \\ &\to \eta_5/(1-\eta_4)\ \sup_{s \ge tf+1}(1+q_s), \quad t \to \infty. \end{aligned}$$

Therefore

$$||e_t|| \le \eta_3(1+q_{tf})tf^{-\eta_4}(1/t)^{1-\eta_4} + \eta_5/(1-\eta_4)\sup_{t \ge tf+1}(1+q_t)$$

as long as $||e_t|| \le \delta$.

Thus since

$$P_{tf}^{-1} = block\ diag(S_0^{-1}, 0_{r\times r}) + \sum_{i=1}^{tf} C_i^T R_i^{-1} C_i$$

then use of Eq. (4.151) gives

$$\begin{aligned} ||e_t|| &\le \eta_6 ||P_{tf}^{-1}||\ ||e_{tf}|| tf^{-\eta_4}(1/t)^{1-\eta_4} + \eta_7 \sup_{t\ge tf+1}||\xi_t|| \\ &\le \eta_6 tf^{-\eta_4}\left(||S_0^{-1}|| + ||\sum_{i=1}^{tf} C_i^T R_i^{-1} C_i|| \right) \vartheta(e_0^\beta, \xi_t) \\ &\quad \times (1+\tilde{\nu}\rho)^{tf-1}(1/t)^{1-\eta_4} + \eta_7 \sup_{t\ge tf+1}||\xi_t||, \end{aligned} \tag{4.193}$$

where

$$\vartheta(e_0^\beta, \xi_t) = (\chi||e_0^\beta|| + \varsigma \max_{t=1,2,\ldots,tf}||\xi_t||),\ \ \upsilon(\rho)\le\tilde{\nu}\rho,\ \ \tilde{\nu}>0,\ \ \rho\to 0.$$

$$\eta_6 = \eta_3(1+q_{tf})\ \eta_7 = \eta_5/(1-\eta_4)\sup_{t\ge tf+1}(1+q_t).$$

We assume without loss of generality that $\rho = ||S_0||$, $tf = ||S_0^{-1}||^{1/\eta_4}$. Then from Eq. (4.193) it follows the existence of $\eta_8 > 0$ and $\eta_9 > 0$ such that if $||e_0^\beta|| < \eta_8$, $\sup_{t\in T}||\xi_t|| < \eta_9$ then $||e_t|| \le \delta$ for all $t\in T$. Therefore, the solution of Eq. (4.99) $x_t = x^*$ is stable under the action of the disturbance ξ_t and if $\xi_t = 0$ then it is asymptotically stable.

4.5 ITERATIVE VERSIONS OF DIFFUSE TRAINING ALGORITHMS

For the diffuse training algorithms we present their iterative modifications oriented towards batch processing, namely, for the DTA from Theorem 4.1:

$$x_t^i = x_{t-1}^i + K_t^i(y_t - h_t(x_{t-1}^i)),\quad t=1,2,\ldots,N,\ \ i=1,2,\ldots,M, \tag{4.194}$$

where $K_t^i = ((K_t^{\beta i})^T, (K_t^{\alpha i})^T)^T$, $h_t(x_{t-1}) = \Phi(z_t, \beta_{t-1})\alpha_{t-1}$,

$$K_t^{\beta i} = (S_t^i + V_t^i(W_t^i)^+(V_t^i)^T)(C_t^\beta)^T + V_t^i(W_t^i)^+(C_t^\alpha)^T, \tag{4.195}$$

$$K_t^{\beta i} = (W_t^i)^+(V_t^i)^T(C_t^\beta)^T + (W_t^i)^+(C_t^\alpha)^T, \tag{4.196}$$

$$S_t^i = S_{t-1}^i/\lambda - S_{t-1}^i(C_t^\beta)^T(\lambda I_m + C_t^\beta S_{t-1}^i(C_t^\beta)^T)^{-1}C_t^\beta S_{t-1}^i/\lambda, \tag{4.197}$$

$$V_t^i = (I_l - S_{t-1}^i (C_t^\beta)^T (\lambda I_m + C_t^\beta S_{t-1}^i (C_t^\beta)^T)^{-1} C_t^\beta) V_{t-1}^i \\ - S_{t-1}^i (C_t^\beta)^T (\lambda I_m + C_t^\beta S_{t-1}^i (C_t^\beta)^T)^{-1} C_t^\alpha, \tag{4.198}$$

$$W_t^i = \lambda W_{t-1}^i + \lambda (C_t^\beta V_{t-1}^i + C_t^\alpha)^T (\lambda I_m + C_t^\beta S_{t-1}^i (C_t^\beta)^T)^{-1} (C_t^\beta V_{t-1}^i + C_t^\alpha), \tag{4.199}$$

$$C_t^\beta = C_t^\beta (x_{t-1}^i) = \partial [\Phi(z_t, \beta_{t-1}^i) \alpha_{t-1}^i] / \partial \beta_{t-1}^i, \quad C_t^\alpha = C_t^\alpha (x_{t-1}^i) = \Phi(z_t, \beta_{t-1}^i), \tag{4.200}$$

with the initialization

$$x_0^1 = (\overline{\beta}^T, 0_{1\times r})^T, \quad S_0^1 = \overline{P}_\beta, \quad V_0^1 = 0_{l\times r}, \quad W_0^1 = 0_{r\times r}, \tag{4.201}$$

$$x_0^i = x_N^{i-1}, \quad S_0^i = S_N^{i-1}, \quad V_0^i = V_N^{i-1}, \quad W_0^i = W_N^{i-1}, \quad i = 2, 3, \ldots, M, \tag{4.202}$$

where i is the number of an iteration, M is the quantity of iterations;

for the DTA Eqs. (4.68)–(4.73):

$$x_t^i = x_{t-1}^i + K_t^i (y_t - h_t(x_{t-1}^i)), \quad t = 1, 2, \ldots, N, \quad i = 1, 2, \ldots, M, \tag{4.203}$$

where $K_t^i = ((K_t^{\beta i})^T, (K_t^{\alpha i})^T)^T, \ h_t(x_{t-1}) = \Phi(z_t, \beta_{t-1}) \alpha_{t-1},$

$$K_t^{\beta i} = S_t^i (C_t^\beta)^T, \quad K_t^{\beta i} = (W_t^i)^+ (C_t^\alpha)^T, \tag{4.204}$$

$$S_t^i = S_{t-1}^i / \lambda - S_{t-1}^i (C_t^\beta)^T (\lambda I_m + C_t^\beta S_{t-1}^i (C_t^\beta)^T)^{-1} C_t^\beta S_{t-1}^i / \lambda, \tag{4.205}$$

$$W_t^i = \lambda W_{t-1}^i + \lambda (C_t^\alpha)^T (\lambda I_m + C_t^\beta S_{t-1}^i (C_t^\beta)^T)^{-1} C_t^\beta V_{t-1}^i \tag{4.206}$$

with the initialization

$$x_0^1 = (\overline{\beta}^T, 0_{1\times r})^T, \quad S_0^1 = \overline{P}_\beta, \quad W_0^1 = 0_{r\times r}, \tag{4.207}$$

$$x_0^i = x_N^{i-1}, \quad S_0^i = S_N^{i-1}, \quad W_0^i = W_N^{i-1}, \quad i = 2, 3, \ldots, M. \tag{4.208}$$

Standard recommendations on the choice of λ are as follows. At the initial stage of training, the value of λ can be sufficiently small to provide a high convergence rate. Then it is offered to gradually increase it to 1 to achieve the necessary accuracy of the obtained solution. One of the variants of choosing λ which provides convergence is proposed in Ref. [76], namely,

$$0 \leq 1-(\lambda_i)^N \leq c/i, \quad c>0, \quad i=1,2,\ldots. \tag{4.209}$$

Based on this result, we will give the convergence conditions for the iterative algorithm proposed in this section. Assume that

(1) for some $L>0$ the Lipschitz condition

$$||\nabla g_t(x)g_t(x)-\nabla g_t(y)g_t(y)|| \leq L||x-y||, \quad \forall x,y\in R^{l+r}, \quad t=1,\ldots,N,$$

is satisfied, where $g_t(x)=y_t-h_t(x)$ and $\nabla g_t(x)$ is the $(l+r)\times m$ matrix of gradients $g_t(x)$;

(2) a sequence of vectors x_t^i, $t=1,\ldots,N$, $i=1,2\ldots$ is bounded;

(3) there is a vector x_t^i such that $W_t=W_t(x_t^i)>0$.

Theorem 4.8.

Under condition Eq. (4.209) and the conditions of items 1–3, the sequence

$$J_t(x_N^i)=\sum_{k=1}^N (y_k-h_t(x_N^i))^T(y_k-h_t(x_N^i))$$

converges and each limiting point of the sequence $\{x_N^i\}$ is a stationary point of $J_t(x)$.

The statement follows from Ref. [76, Proposition 2] if

$$P_t=P_t(x_t^{i-1})=\begin{pmatrix} S_t^i+V_t^i(W_t^i)^{-1}(V_t^i)^T & V_t^i(W_t^i)^{-1} \\ (W_t^i)^{-1}(V_t^i)^T & (W_t^i)^{-1}\end{pmatrix}>0$$

and from the representation

$$|P_t|=|S_t^i+V_t^i(W_t^i)^{-1}(V_t^i)^T|/|W_t^i+V_t^i(S_t^i)^{-1}(V_t^i)^T|$$

which was established in the proof of Theorem 4.7.

4.6 DIFFUSE TRAINING ALGORITHM OF RECURRENT NEURAL NETWORK

Consider the recurrent neural network which is described by the state-space model

$$p_t=\begin{pmatrix}\sigma(a_1^Tp_{t-1}+b_1^Tz_t+d_1)\\ \cdots \\ \sigma(a_q^Tp_{t-1}+b_q^Tz_t+d_q)\end{pmatrix}=\sigma(Ap_{t-1}+Bz_t+d), \tag{4.210}$$

$$y_t = \begin{pmatrix} c_1^T p_t \\ \dots \\ c_m^T p_t \end{pmatrix} + \xi_t = Cp_t + \xi_t, \quad t = 1, 2, \ldots, N, \tag{4.211}$$

where $p_t \in R^q$ is a state vector, $z_t \in R^n$ is a vector of inputs, $y_t \in R^m$ is a vector of outputs, $a_i \in R^q$, $b_i \in R^n$, $d_i \in R^1$, $c_i \in R^q$, $i = 1, 2, \ldots, q$ are vectors of unknown parameters, $\sigma(x)$ is the sigmoid function, $A = (a_1, a_2, \ldots, a_q) \in R^{q \times q}$, $B = (b_1, b_2, \ldots, b_n) \in R^{q \times n}$, $d = (d_1, d_2, \ldots, d_q) \in R^q$, $\xi_t \in R^m$ is a random process which has uncorrelated values, zero expectation and a covariance matrix R_t.

Let us turn to a system with more convenient parameterization for us including Eqs. (4.210) and (4.211) as a special case

$$p_t = \Psi(z_t, p_{t-1}, \beta), \tag{4.212}$$

$$y_t = \begin{Bmatrix} p_t^T & 0 & \cdots & 0 \\ & \cdots & & \\ 0 & 0 & \cdots & p_t^T \end{Bmatrix} \alpha + \xi_t = \Phi(p_t)\alpha + \xi_t, \quad t = 1, 2, \ldots, N, \tag{4.213}$$

where

$$\alpha = (\alpha_1, \alpha_2, \ldots, \alpha_{qm})^T = (c_1^T, c_2^T, \ldots, c_m^T) \in R^{qm},$$

$$\beta = (\beta_1, \beta_2, \ldots, \beta_{qm+q})^T = (a_1^T, a_2^T, \ldots, a_q^T, b_1^T, b_2^T, \ldots, b_q^T, d^T) \in R^{qm+q}$$

are vectors of unknown parameters including elements of A, B, C, d, $\Psi(z_t, p_{t-1}, \beta) \in R^q$ is a vector of given nonlinear functions, $\Phi(p_t) \in R^k, k = q^2 + mq + q$.

We suppose that the unknown parameters in Eqs. (4.212) and (4.213) satisfy conditions A1, A2, and A3 and the quality criteria at a moment t are defined by Eq. (4.4).

From Eqs. (4.212) and (4.213) follow relations for matrix functions $C_t^{\alpha} \in R^{m \times qm}$, $C_t^{\beta} \in R^{m \times pm}$ included in the description of the DTA:

$$C_t^{\alpha} = \begin{Bmatrix} p_t^T & 0 & \dots & 0 \\ & \dots & & \\ 0 & 0 & \dots & p_t^T \end{Bmatrix}, \tag{4.214}$$

$$C_t^{\beta} = \partial\Phi(p_t)/\partial\beta_{t-1}\alpha_{t-1} = \partial\Phi(p_t)/\partial p_t \partial p_t/\partial\beta_{t-1}\alpha_{t-1}, \tag{4.215}$$

where

$$\partial p_t/\partial\beta_{t-1} = \partial\Psi(z_t,p_{t-1},\beta_{t-1})/\partial p_{t-1}\partial p_{t-1}/\partial\beta_{t-1} + \partial\Psi(z_t,p_{t-1},\beta_{t-1})/\partial\beta_{t-1}, \tag{4.216}$$

$$\partial p_0/\partial\beta_0 = 0.$$

After completion of the training phase the system Eq. (4.210) can be used to predict the behavior of the output y_t. At the same time, the prediction should begin immediately after the completion of training, that is, from the moment of $t = N+1$ with the initial condition p_N. In Chapter 5, Diffuse Kalman Filter, a more efficient training algorithm free from this disadvantage based on diffuse Kalman filter will be proposed.

4.7 ANALYSIS OF TRAINING ALGORITHMS WITH SMALL NOISE MEASUREMENTS

Let us consider the behavior of the proposed training algorithms under a small measurement noise whose intensity matrix R_t enters in algorithms through the inverse matrix $N_{1t}^{-1} = (C_t^\beta S_t(C_t^\beta)^T + R_t)^{-1}$. Taking this into account it is important to understand their behavior as $R_t \to 0$.

We first consider the DTA described by expressions (4.58)–(4.64).

Theorem 4.9.
For a given N, $R_t = \varepsilon I_m$, $\varepsilon \to 0$, $\varepsilon > 0$, and $S_0 > 0$

$$\tilde{K}_t = \lim_{\varepsilon\to 0} K_t^{dif} = \begin{pmatrix} Q_t^+(C_t^\beta)^T \\ 0_{r\times l} \end{pmatrix} + \begin{pmatrix} G_t \\ I_r \end{pmatrix} L_t^+ \begin{pmatrix} G_t \\ I_r \end{pmatrix}^T \begin{pmatrix} (C_t^\beta)^T \\ (C_t^\alpha)^T \end{pmatrix}, \tag{4.217}$$

where

$$Q_t = Q_{t-1} + (C_t^\beta)^T C_t^\beta, \quad Q_0 = 0_{l\times l}, \tag{4.218}$$

$$G_t = (I_l - Q_t^+(C_t^\beta)^T C_t^\beta)G_{t-1} - Q_t(C_t^\beta)^T C_t^\alpha, \quad G_0 = 0_{l\times r}, \tag{4.219}$$

$$L_t = L_{t-1} + (C_t^\beta G_t + C_t^\alpha)^T \overline{H}_t (C_t^\beta G_t + C_t^\alpha), L_0 = 0_{r\times r}. \tag{4.220}$$

$$\overline{H}_t = (I_l - \tilde{H}_t\tilde{H}_t^+)(I_l + C_t^\beta Q_t^+(C_t^\beta)^T)^{-1}, \tag{4.221}$$

$$\tilde{H}_t = (I_l + C_t^{\beta} Q_{t-1}^{+} (C_t^{\beta})^T)^{-1} C_t^{\beta} H_t (C_t^{\beta})^T, \tag{4.222}$$

$$H_t = S_0(I_l - Q_{t-1} Q_{t-1}^{+}), \quad t = 1, 2, \ldots, N. \tag{4.223}$$

Proof

Since

$$S_t^{-1} = S_0^{-1} + \sum_{k=1}^{t} (C_k^{\beta})^T C_k^{\beta} / \varepsilon = \left(S_0^{-1} \varepsilon + \sum_{k=1}^{t} (C_k^{\beta})^T C_k^{\beta} \right) / \varepsilon, \quad t = 1, 2, \ldots, N$$

then Lemma 2.1 implies

$$\begin{aligned} S_t &= \left(S_0^{-1} / \varepsilon + \sum_{k=1}^{t} (C_k^{\beta})^T C_k^{\beta} \right) \varepsilon \\ &= S_0(I_l - Q_t Q_t^{+}) + Q_t^{+} \varepsilon + O(\varepsilon^2), \quad t = 1, 2, \ldots, N \end{aligned} \tag{4.224}$$

if we put $\Omega_1 = S_0^{-1}$, $F_t = C_t^{\beta}$.

Using Lemma 2.2 and Eq. (2.24), we obtain

$$\lim_{\varepsilon \to 0} S_t (C_t^{\beta})^T / \varepsilon = Q_t^{+} (C_t^{\beta})^T.$$

We show first that

$$S_{t-1} (C_t^{\beta})^T (R_t + C_t^{\beta} S_{t-1} (C_t^{\beta})^T)^{-1} = S_t (C_t^{\beta})^T R_t^{-1}.$$

Multiplying Eq. (4.61) on the right $(C_t^{\beta})^T R_t^{-1}$ gives

$$\begin{aligned} S_t (C_t^{\beta})^T R_t^{-1} &= S_{t-1} (C_t^{\beta})^T R_t^{-1} - S_{t-1} (C_t^{\beta})^T (R_t + C_t^{\beta} S_{t-1} (C_t^{\beta})^T)^{-1} \\ &\quad \times C_t^{\beta} S_{t-1} (C_t^{\beta})^T R_t^{-1} \\ &= S_{t-1} (C_t^{\beta})^T (I_l - (R_t + C_t^{\beta} S_{t-1} (C_t^{\beta})^T)^{-1} C_t^{\beta} S_{t-1} (C_t^{\beta})^T) R_t^{-1} \\ &= S_{t-1} (C_t^{\beta})^T (R_t + C_t^{\beta} S_{t-1} (C_t^{\beta})^T)^{-1}. \end{aligned}$$

It follows that

$$\lim_{\varepsilon \to 0} S_{t-1} (C_t^{\beta})^T (I_m \varepsilon + C_t^{\beta} S_{t-1} (C_t^{\beta})^T)^{-1} = \lim_{\varepsilon \to 0} S_t (C_t^{\beta})^T / \varepsilon = Q_t^{+} (C_t^{\beta})^T, \tag{4.225}$$

$$\lim_{\varepsilon \to 0} V_t = G_t. \tag{4.226}$$

Let us find the asymptotic representation for N_t^{-1} as $\varepsilon \to 0$. We transform N_t using Eq. (4.224) as follows

$$N_t = I_m\varepsilon + C_t^\beta S_{t-1}(C_t^\beta)^T = I_m\varepsilon + C_t^\beta(S_0(I_l - Q_{t-1}Q_{t-1}^+)$$
$$+ Q_{t-1}^+\varepsilon + O(\varepsilon^2))(C_t^\beta)^T + O(\varepsilon^2)$$
$$= C_t^\beta H_t(C_t^\beta)^T + (I_m + C_t^\beta Q_{t-1}^+(C_t^\beta)^T)\varepsilon + O(\varepsilon^2)$$
$$= (I_m + C_t^\beta Q_{t-1}^+(C_t^\beta)^T)(\tilde{H}_t + I_m\varepsilon) + O(\varepsilon^2).$$

Using Lemma 2.1 yields

$$(\tilde{H}_t + I_m\varepsilon)^{-1} = (I_l - \tilde{H}_t\tilde{H}_t^+)/\varepsilon + \tilde{H}_t^+ + O(\varepsilon^2).$$

Therefore

$$N_t^{-1} = (I_l - \tilde{H}_t\tilde{H}_t^+)(I_m + C_t^\beta Q_{t-1}^+(C_t^\beta)^T)^{-1}/\varepsilon + O(1) = \overline{H}_t/\varepsilon + O(1).$$

We have

$$W_t = W_{t-1} + (C_t^\beta V_{t-1} + C_t^\alpha)^T(R_t + C_t^\beta S_{t-1}(C_t^\beta)^T)^{-1}(C_t^\beta V_{t-1} + C_t^\alpha)$$
$$= W_{t-1} + (C_t^\beta V_{t-1} + C_t^\alpha)^T N_t^{-1}(C_t^\beta V_{t-1} + C_t^\alpha)$$
$$= W_{t-1} + (C_t^\beta V_{t-1} + C_t^\alpha)^T(\overline{H}_t + O(\varepsilon))(C_t^\beta V_{t-1} + C_t^\alpha)/\varepsilon.$$

This implies that

$$L_t = \lim_{\varepsilon\to 0} W_t\varepsilon = \sum_{k=1}^{t}(C_k^\beta G_{k-1} + C_k^\alpha)^T\overline{H}_k(C_k^\beta G_{k-1} + C_t^\alpha),$$

$$\lim_{\varepsilon\to 0}\begin{pmatrix} V_t \\ I_r \end{pmatrix} W_t^+ \begin{pmatrix} V_t \\ I_r \end{pmatrix}^T \begin{pmatrix} (C_t^\beta)^T \\ (C_t^\alpha)^T \end{pmatrix} R_t^{-1} = \begin{pmatrix} G_t \\ I_r \end{pmatrix} L_t^+ \begin{pmatrix} G_t \\ I_r \end{pmatrix}^T \begin{pmatrix} (C_t^\beta)^T \\ (C_t^\alpha)^T \end{pmatrix}.$$

A similar result holds for the two-stage algorithm described by Eqs. (4.58) and (4.71)–(4.73).

Theorem 4.10.
For a given N, $R_t = \varepsilon I_m$, $\varepsilon \to 0$, $\varepsilon > 0$, and $S_0 > 0$

$$\overline{K}_t^\beta = \lim_{\varepsilon\to 0} K_t^{dif,\beta} = Q_t^+(C_t^\beta)^T,\ \ \overline{K}_t^\alpha = \lim_{\varepsilon\to 0} K_t^{dif,\alpha} = L_e^+(C_t^\alpha)^T, \tag{4.227}$$

where

$$Q_t = Q_{t-1} + (C_t^{\beta})^T C_t^{\beta}, \quad Q_0 = 0_{l\times l}, \tag{4.228}$$

$$L_t = L_{t-1} + (C_t^{\alpha})^T \overline{H}_t C_t^{\alpha}), \quad L_0 = 0_{r\times r}. \tag{4.229}$$

The statement is proved by the same scheme as the previous Theorem 4.9.

4.8 EXAMPLES OF APPLICATION

4.8.1 Identification of Nonlinear Static Plants

Suppose that we want to describe the plant model with the scalar output using a multilayer perceptron with one hidden layer and linear output activation function

$$y_t = \sum_{k=1}^{p} w_k \sigma\left(\sum_{j=1}^{q} a_{kj} z_{jt} + b_k\right) + \xi_t, \quad t = 1, 2, \ldots, N, \tag{4.230}$$

where z_{jt}, $j = 1, 2, \ldots, q$ are inputs of the plant, y_t is the output of the plant, $a_{kj}, b_k, k = 1, 2, \ldots, p$, $j = 1, 2, \ldots, q$ are weights and biases of the hidden layer, respectively, w_k, $k = 1, 2, \ldots, p$ are weights of the output layer, $\sigma(x) = (1 + \exp(-x))^{-1}$ is the hidden layer activation function, and ξ_t is centered random process with uncorrelated values and variance R_t.

It is required to estimate neuron network weights and biases from observations of input–output pairs $\{z_i, y_i\}$, $t = 1, 2, \ldots, N$ using the DTA.

Let us rewrite Eq. (4.230) in a form more suitable for us Eq. (4.1)

$$y_t = \Phi(z_t, \beta)\alpha + \xi_t, \quad t = 1, 2, \ldots, N, \tag{4.231}$$

where

$$\beta = (a_{11}, a_{12}, \ldots, a_{1q}, b_1, a_{21}, a_{22}, \ldots, a_{2q}, b_2, \ldots, a_{p1}, a_{p2}, \ldots, a_{pq}, b_p)^T \in R^{(q+1)p},$$

$$\alpha = (w_1, w_2, \ldots, w_p)^T \in R^p, \quad \Phi(z_t, \beta)\alpha = \sum_{k=1}^{p} w_k \sigma\left(\sum_{j=1}^{q} a_{kj} z_{jt} + b_k\right).$$

To use the DTA we have to find:

$$C_t^{\beta} = \partial[\Phi(z_t, \beta)\alpha]/\partial\beta, \quad C_t^{\alpha} = \Phi(z_t, \beta_{t-1}) \tag{4.232}$$

and select initial values β_0 and S_0.

From Eq. (4.231) it follows that

$$C_t^{\alpha} = (\sigma(a_1 z_t + b_1), \sigma(a_2 z_t + b_2), \ldots, \sigma(a_r z_t + b_r)) \in R^{1\times p}, \qquad (4.233)$$

$$C_t^{\beta} = (\tilde{a}_{11}, \tilde{a}_{12}, \ldots, \tilde{a}_{1q}, \tilde{b}_1, \tilde{a}_{21}, \tilde{a}_{22}, \ldots, \tilde{a}_{2q}, \tilde{b}_2, \ldots, \tilde{a}_{p1}, \tilde{a}_{p2}, \ldots, \tilde{a}_{pq}, \leftrightarrow b_p)^T \in R^{(q+1)p}, \qquad (4.234)$$

where

$$z_t = (z_{1t}, z_{2t}, \ldots, z_{qt}, 1)^T \in R^{q+1}, \quad a_k = (a_{k1}, a_{k2}, \ldots, a_{kq}, b_k), \quad k = 1, 2, \ldots, p,$$

$$\tilde{a}_{kj} = \frac{\partial}{\partial a_{kj}} \left[\sum_{k=1}^{p} w_k \sigma \left(\sum_{j=1}^{q} a_{kj} z_{jt} + b_k \right) \right] = \sum_{k=1}^{p} w_k \frac{d}{dx} \sigma(x) z_{jt},$$

$$\tilde{b}_k = \frac{\partial}{\partial b_k} \left[\sum_{k=1}^{p} w_k \sigma \left(\sum_{j=1}^{q} a_{kj} z_{jt} + b_k \right) \right] = \sum_{k=1}^{p} w_k \frac{d}{dx} \sigma(x),$$

$$x = \sum_{j=1}^{q} a_{kj} z_{jt} + b_k.$$

We choose a small random vector $\overline{\beta}$ with a covariance matrix $\overline{P}_{\beta}$ to initialize β_t and set

$$\beta_0 = \overline{\beta}, \quad S_0 = h\overline{P}_{\beta}, \quad h \in (0, 1].$$

Example 4.1.

Let us illustrate the efficiency of the results obtained in this chapter by one of the standard examples used in testing algorithms [7]. Consider the problem of approximation of the function that is described by the expression

$$y(x) = \begin{cases} \sin(x)/x, & x \neq 0 \\ 1 & x = 0. \end{cases}$$

Training and testing sets (x_i, y_i) include 2000 points each and values of x_i are uniformly distributed on the interval $[-10, 10]$. Noise uniformly distributed on the interval [0.2, 0.2] is added to training values y_i and it is assumed that the testing set does not contain any noise. The sigmoid AF is used and the approximation accuracy of the ELM is compared with the accuracy of the diffuse algorithms with sequential processing. As a measure of accuracy, an estimate of the 90th percentile of the approximation error obtained from 500 samples is used. Weights and biases of the hidden layer are assumed to be random quantities uniformly distributed on the intervals [1, 1] and [0, 1], respectively, $R_t = 0.16/12$. Introduce the following notations: DTA1 is described by the relations Eqs. (2.74)–(2.76) which estimate output layer parameters only (the diffuse version of ELM algorithms), DTA2 and DTA3 are described by the relations (4.58)–(4.64) and (4.58), (4.71)–(4.73), respectively. The modeling results are presented in Table 4.1. As is easily seen, the DTAs surpass the ELM in accuracy for both training and testing sets. DTA2 and DTA3 yield practically undistinguishable results concerning accuracy but DTA3 surpasses DTA2 in speed by 32%. Fig. 4.1 presents the curves of approximating dependencies 1 and 2 (ELM and DTA2, respectively) for five neurons of the hidden layer and the used training set.

Table 4.1 Training errors

Number of neurons in the hidden layer	Accuracy of algorithms					
	In training			In testing		
	ELM	DTA1	DTA2	ELM	DTA1	DTA2
4	0.315	0.169	0.174	0.295	0.125	0.13
5	0.272	0.159	0.159	0.246	0.108	0.11
6	0.205	0.148	0.147	0.17	0.092	0.092

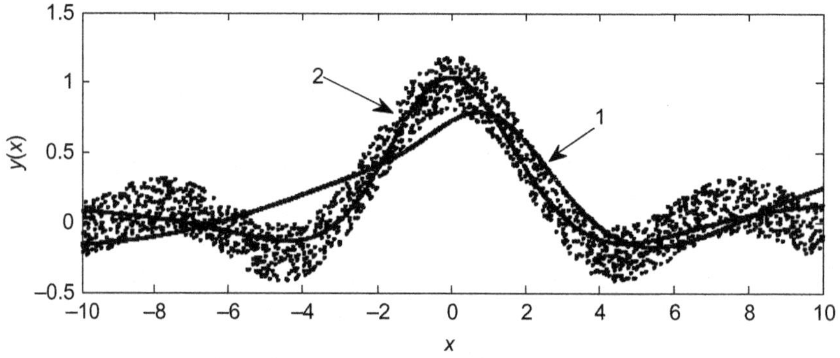

Figure 4.1 Approximating dependencies for DTA1 (1) and DTA2 (2).

Let us now assume that a plant model is approximated with help of a neuro-fuzzy Sugeno network of zero order with Gaussians for membership function (MF)

$$y_t = \Phi(z_t, \beta)\alpha + \xi_t = \frac{\sum_{i=1}^{q} \alpha_i \prod_{j=1}^{n} \mu_{ij}(z_{jt}, m_{ij}, \sigma_{ij})}{\sum_{i=1}^{q} \prod_{j=1}^{n} \mu_{ij}(z_{jt}, m_{ij}, \sigma_{ij})} + \xi_t, \quad t = 1, 2, \ldots, N, \tag{4.235}$$

where

$$\mu_{ij}(z_{jt}, m_{ij}, \sigma_{ij}) = \exp(-(z_{jt} - m_{lj})^2 / 2\sigma_{lj}^2), \quad i = 1, 2, \ldots, q, j = 1, 2, \ldots, n, \tag{4.236}$$

$$\begin{aligned} &\alpha = (\alpha_1, \alpha_2, \ldots, \alpha_q)^T \in R^q, \\ &\beta = \big(m_{11}, m_{12}, \ldots, m_{1n}, m_{21}, m_{22}, \ldots, m_{2n}, \ldots, m_{q1}, m_{q2}, \ldots, m_{qn}, \\ &\qquad \sigma_{11}, \sigma_{12}, \ldots, \sigma_{1n}, \sigma_{21}, \sigma_{22}, \ldots, \sigma_{2n}, \ldots, \sigma_{q1}, \sigma_{q2}, \ldots, \sigma_{qn}\big)^T \in R^{2nq}. \end{aligned} \tag{4.237}$$

It is required to estimate α and β from observations of input–output pairs $\{z_i, y_i\}$, $t = 1, 2, \ldots, N$ using the diffuse algorithms.

To use these algorithms we have to find the expressions for the matrix functions C_t^β, C_t^α which are included in their description and select initial values β_0 and S_0.

We have

$$\Phi(z_t, \beta)\alpha = \frac{\sum_{i=1}^{q} \alpha_i a_i(z_t, \beta)}{\sum_{i=1}^{q} a_i(z_t, \beta)}, \quad t = 1, 2, \ldots, N, \tag{4.238}$$

where $a_i(z_t, \beta) = \prod_{j=1}^{n} \mu_{ij}(z_{jt}, m_{ij}, \sigma_{ij})$.

It follows from this that

$$C_t^\alpha = \frac{1}{\sum_{i=1}^{q} a_i(z_t, \beta)} (a_1(z_t, \beta), a_2(z_t, \beta), \ldots, a_q(z_t, \beta)) \in R^{1 \times q}, \tag{4.239}$$

$$\tilde{m}_{ij} = \frac{\partial}{\partial m_{ij}} \Phi(z_t, \beta)\alpha = \alpha_i (b(z_t, \beta) - 1)\mu_{ij}(z_{jt}, m_{ij}, \sigma_{ij})(z_{jt} - m_{lj})/\sigma_{lj}^2,$$

$$\tilde{\sigma}_{ij} = \frac{\partial}{\partial \sigma_{ij}} \Phi(z_t, \beta)\alpha$$
$$= \alpha_i(b(z_t, \beta) - 1)\mu_{ij}(z_{jt}, m_{ij}, \sigma_{ij})(z_{jt} - m_{lj})^2 / \sigma_{lj}^3, \; i = 1,2,\ldots,q, \; j = 1,2,\ldots,n,$$

$$C_t^{\beta} = \left(\tilde{m}_{11}, \tilde{m}_{12}, \ldots, \tilde{m}_{1n}, \tilde{m}_{21}, \tilde{m}_{22}, \ldots, \tilde{m}_{2n}, \ldots, \tilde{m}_{q1}, \tilde{m}_{q2}, \ldots, \tilde{m}_{qn},\right.$$
$$\left.\tilde{\sigma}_{11}, \tilde{\sigma}_{12}, \ldots, \tilde{\sigma}_{1n}, \tilde{\sigma}_{21}, \tilde{\sigma}_{22}, \ldots, \tilde{\sigma}_{2n}, \ldots, \tilde{\sigma}_{q1}, \tilde{\sigma}_{q2}, \ldots, \tilde{\sigma}_{qn}\right)^T \in R^{2nq}, \tag{4.240}$$

where $b(z_t, \beta) = \sum_{i=1}^{q} \alpha_i a_i(z_t, \beta)$.

For the parameters of the MF often there is some expert knowledge which is connected with specifics of the decided problem. As an example for Gaussians Eq. (4.236), values $\overline{m}_{lj}$ and $\overline{\sigma}_{lj}$ can be specified before training. A natural requirement after optimization is to keep the original order of linguistic values $m_{lj+1} > m_{lj}$ and $\sigma_{lj} > 0$ imposes the restriction on the size of their spread with respect to the selected values $\overline{m}_{lj}$ and $\overline{\sigma}_{lj}$. Taking this into account, the MF parameters before training are interpreted as random values with expectations $\overline{m}_{lj}$ and $\overline{\sigma}_{lj}$ and sufficiently small variances

$$S_{\beta} = E((\beta - m_{\beta})(\beta - m_{\beta})^T) = diag(s_1, \ldots, s_k),$$

where $s_i \in [0, \varepsilon]$ are uniformly distributed numbers, $\varepsilon > 0$ is a small parameter. Similar considerations are valid for any parameterized MF. In contrast to β, in respect to α there is no a priori information in respect to α.

Example 4.2.

Let us consider the problem of identification of a static object from input–output data [77] (represented by points in Fig. 4.2) with the help of the zero-order Sugeno fuzzy inference system. There are two peaks observed against the background of a decaying trend and broadband noise. Six membership functions and $n = 18$ unknown parameters were used; $q = 12$ of these parameters are in the description of the MFs. Under the assumption that the MFs are Gaussian, it is easy to determine the matrix C_t from the expressions (4.239) and (4.240). The centers and width of MFs were specified using the standard methodology of initial (before optimization) arrangements of MF parameters [78] (Fig. 4.3), and the deviations from them were assumed to be equal to 5 and 0.01, respectively. Fig. 4.4 presents dependencies of the mean-square estimation error

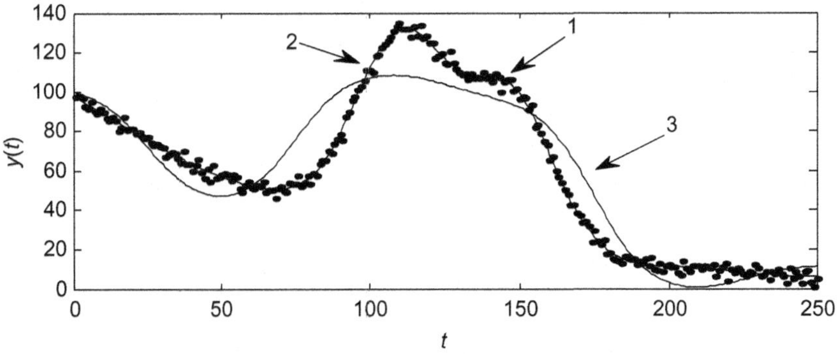

Figure 4.2 Dependencies of the plant and model outputs on the number of observations.

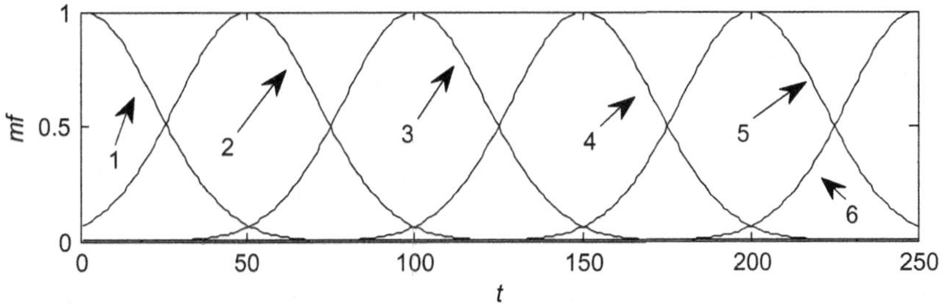

Figure 4.3 Dependencies of six membership functions on the number of observations before optimization.

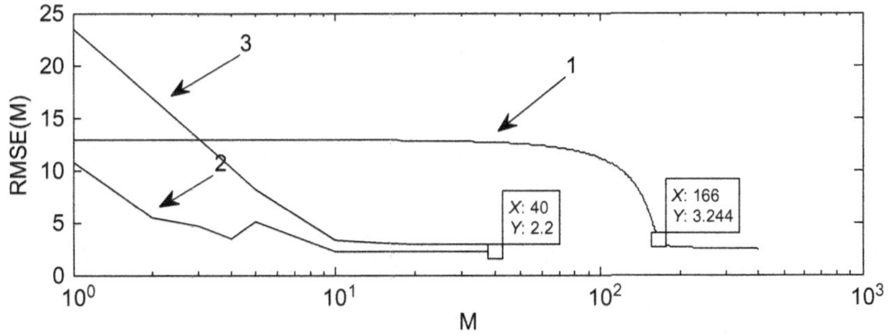

Figure 4.4 Dependencies of the mean-square estimation errors on the number of iterations *M*.

on the number of iterations $RMSE(M)$ (epochs) with the forgetting parameter $\lambda_i = \max\{1 - 0.05/i, 0.99\}$, $i = 1, 2, ..., M$. The plots of curves 1, 2, and 3 are constructed with the help of the hybrid algorithm of the ANFIS system, the iterative DTAs that are described by the systems

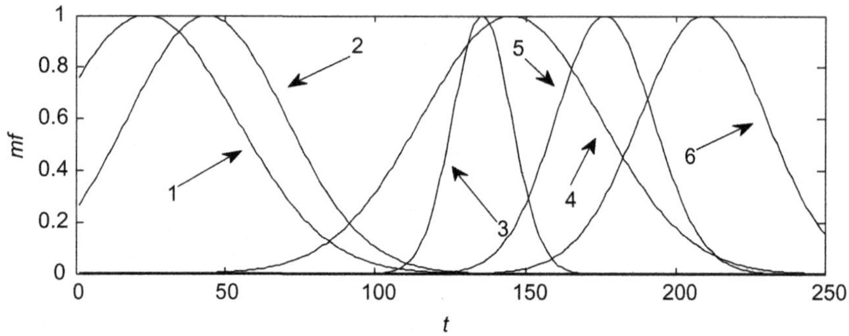

Figure 4.5 Dependencies of six membership functions on the number of observations after optimization.

Eqs. (4.194)–(4.202) and (4.203)–(4.208), respectively. It is seen that the proposed algorithms considerably exceed the hybrid algorithm of the ANFIS system in convergence speed.

It should be noted that the forgetting parameter exerts a considerable influence on the convergence rate of an algorithm. In fact, when $\lambda_i = 1$ and the number of iterations equals 40, we have $RMSE(40) = 6.8$. In Figs. 4.3 and 4.5 the membership functions obtained before and after optimization are shown, respectively. Despite considerable changes in the MFs after optimization, the initial linguistic order has remained in these figures.

4.8.2 Identification of Nonlinear Dynamic Plants

Example 4.3.
Consider the problem of the plant identification described by the nonlinear difference equation [58]

$$y_t = y_{t-1}y_{t-2}(y_{t-1} + 2.5)/(1 + y_{t-1}^2 + y_{t-2}^2) + u_{t-1}.$$

Values of the training set of the input u_t are uniformly distributed over the interval $[-2, 2]$ and values of the test sample are specified by the expression $u_t = \sin(2\pi t/250)$. The model of the object is found in the form of a nonlinear autoregression moving average model

$$y_t = f(y_{t-1}, y_{t-2}, u_{t-1}),$$

where f (.) is a multilayer perceptron with one hidden layer and linear output activation function.

In contrast to Section 2.3.1 we will simultaneously estimate all parameters included in the description of the perceptron. The initial values of the weights of the hidden layer and biases are selected from the uniform distribution on the intervals $[-1, 1]$ and $[0, 1]$, respectively. Let us introduce the following notations: DTA1 are described by the relations (4.35)−(4.41), DTA2 is described by the relations Eqs. (2.30) and (2.31) which estimate output layer parameters only (the diffuse version of ELM algorithms). Table 4.2 shows the values of the 90th percentile of the approximation output error DTA1. For comparison, in the same table, the simulation results are given in the evaluation of the output layer weights only (DTA2). Fig. 4.6 shows the system outputs and models with five neurons in the hidden layer. Here curves 1 and 2 are outputs of DTA2 and DTA1, respectively, and curve 3 is the output of the plant.

Table 4.2 Training errors

Sizes of training *N* and testing *M* sets	**Number of neurons in the hidden layer**	**Training**		**Testing**	
		DTA1	**DTA2**	**DTA1**	**DTA2**
$N=2000$,	5	0.13	0.57	0.06	0.59
$M=500$	10	0.12	0.23	0.05	0.18
	15	0.13	0.14	0.05	0.08
	20	0.13	0.12	0.05	0.04
$N=500$,	4	0.16	0.83	0.11	0.83
$M=500$	5	0.15	0.58	0.1	0.58
	6	0.14	0.48	0.1	0.46
	10	0.14	0.23	0.09	0.19

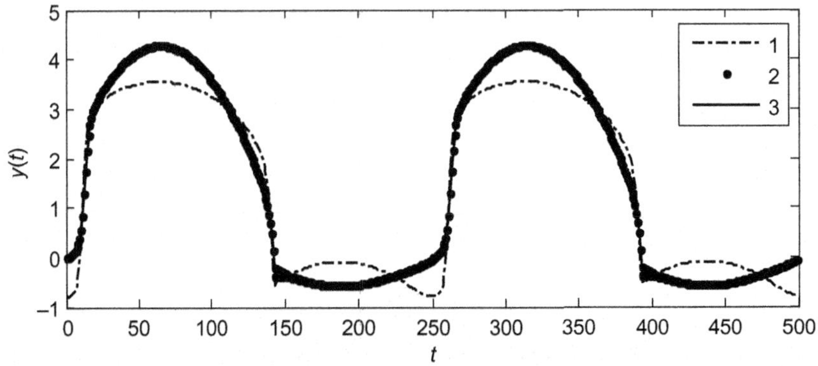

Figure 4.6 Dependencies of the plant output and the models outputs on the time.

Example 4.4.

Consider the problem of identification of two reservoir capacity using the input–output data [79]. Voltage (an input) is applied to pump fluid which fills the upper tank. From the upper tank it flows through a hole in the bottom container. The fluid level in the lower container defines a output of the system $y(t)$. The data contain 3000 values $u(t)$ in volts and $y(t)$ in meters measured with the step of 0.2 s, which are shown in Figs. 4.7 and 4.8, respectively.

The plant model is searched in the form of the multilayer perceptron with one hidden layer and linear output activation function and a vector of inputs

$$z_t = (y_{t-1}, y_{t-2}, y_{t-3}, y_{t-4}, y_{t-5}, u_{t-3})^T.$$

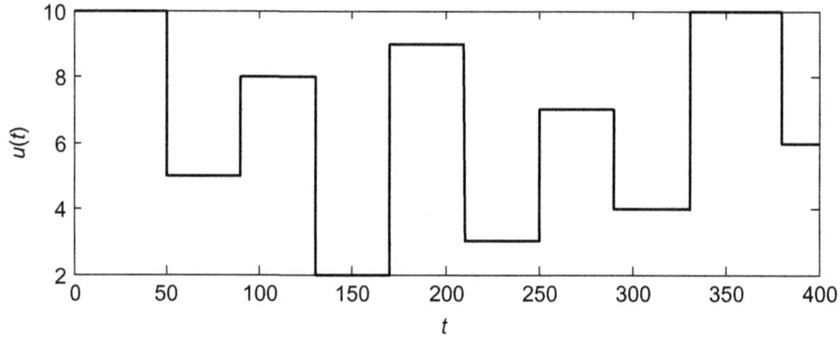

Figure 4.7 Dependency of the plant input on the time.

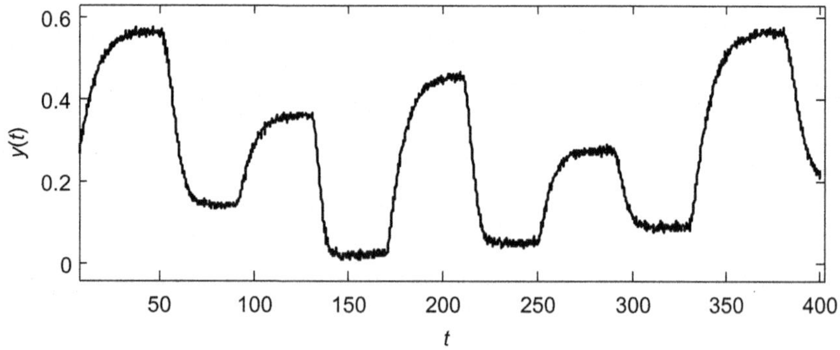

Figure 4.8 Dependency of the plant output on the time.

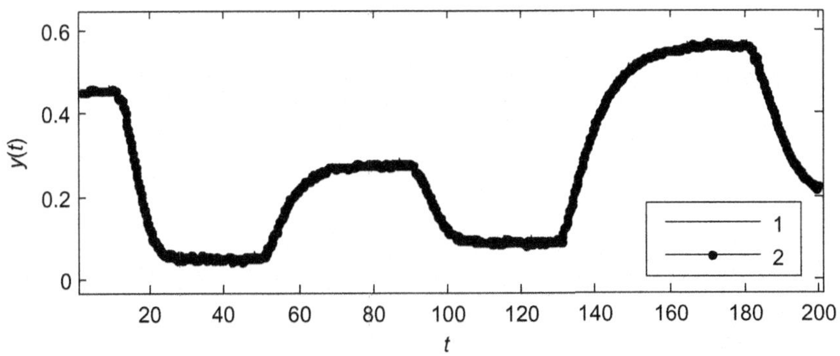

Figure 4.9 Dependencies of the plant output, the model output on the time.

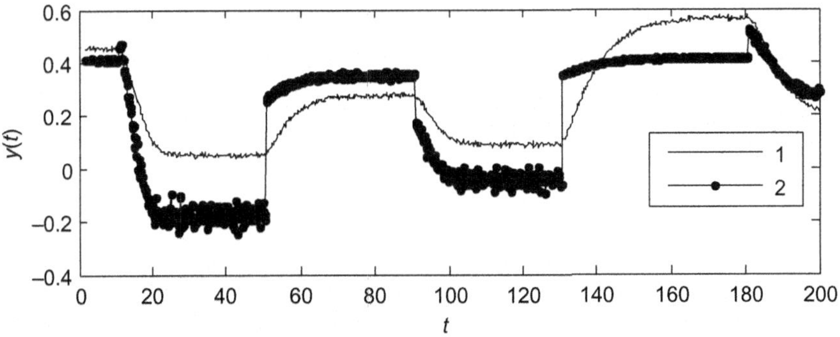

Figure 4.10 Dependencies of the plant output, the model output on the time with output layer estimation parameters only.

The size of the training sample is $N = 1000$, the testing sample is $M = 1000$. Hidden layer weights and bias are selected as in the previous example. The parameter of forgetting is determined by the expression $\lambda_k = \{\max(1 - 0.05/k, 0.99\}$. Fig. 4.9 shows the output values of the plant (curve 1) and its model built using the DTA (curve 2) and the expressions (4.194)–(4.202) with five neurons in the hidden layer with the simultaneous training of all unknown parameters for 20 ages. Fig. 4.10 shows output of the plant (curve 1) and the model built using the DTA (curve 2) of Theorem 2.1 with five neurons in the hidden layer and output layer estimation parameters only.

4.8.3 Example of Classification Task

Let us illustrate use of the iterative DTA to solve the problem of classification of objects according to the training sample $\{z_i, y_i\}$, $t = 1, 2, \ldots, N$, where $z_t = (z_{1t}, z_{2t}, \ldots, z_{nt})^T \in R^n$ is a vector of features characterizing a recognizable object, $y_i \in Y = \{1, 2, \ldots, m\}$ is a finite set of classes numbers.

We use the following form of the RBNN [24]:

$$y_{it} = \sum_{k=1}^{p} w_{ik}\varphi(||z_t - a_k||^2) + w_{i0}, \quad i = 1, 2, \ldots, m, \quad t = 1, 2, \ldots, N, \tag{4.241}$$

where $a_k \in R^n$, $k = 1, 2, \ldots, p$ are centers, w_{ik}, $i = 1, 2, \ldots, m$, $k = 1, 2, \ldots, p$ are weights of the output layer, $w_{i0}, i = 1, 2, \ldots, m$ are biases, and $\varphi(||z_t - a_k||^2) = (||z_t - a_k||^2 + 1)^{-1/2}$ is a basis function.

The network is trained so that its k-th output is equal to 1 and all the others are equal to 0 when the input vector z_i belongs to the class with number k.

Let us turn to the equivalent description

$$y_t = \Phi(z_t, \beta)\alpha, \quad t = 1, 2, \ldots, N, \tag{4.242}$$

where

$$\alpha = (w_{10}, w_{11}, \ldots, w_{1p}, w_{20}, w_{21}, \ldots, w_{2p}, w_{m0}, w_{m1}, \ldots, w_{mp})^T \in R^{(p+1)m},$$

$$\beta = (a_1^T, a_2^T, \ldots, a_p^T)^T \in R^{np},$$

$$\Phi(z_t, \beta) = block\ diag(\Phi_1(z_t, \beta), \Phi_2(z_t, \beta), \ldots, \Phi_m(z_t, \beta)) \in R^{m \times m(p-1)},$$

$$\Phi_i(z_t, \beta) = (1, \varphi(||z_t - a_1||^2), \varphi(||z_t - a_2||^2), \ldots, \varphi(||z_t - a_p||^2)) \in R^{1 \times (p+1)}.$$

Whence follow equations for matrix functions included in the description of the DTA:

$$C_t^\alpha = \begin{pmatrix} 1 & \varphi(||z_t - a_1||^2) & \varphi(||z_t - a_2||^2) & \ldots & \varphi(||z_t - a_p||^2) \\ 1 & \varphi(||z_t - a_1||^2) & \varphi(||z_t - a_2||^2) & \ldots & \varphi(||z_t - a_p||^2) \\ \ldots & \ldots & \ldots & \ldots & \ldots \\ 1 & \varphi(||z_t - a_1||^2) & \varphi(||z_t - a_2||^2) & & \varphi(||z_t - a_p||^2) \end{pmatrix},$$

$$C_t^\beta = \begin{pmatrix} w_{11}\frac{\partial}{\partial a_1}\varphi(||z_t - a_1||^2) & w_{12}\frac{\partial}{\partial a_2}\varphi(||z_t - a_2||^2) & \dots & w_{1p}\frac{\partial}{\partial a_p}\varphi(||z_t - a_p||^2) \\ w_{21}\frac{\partial}{\partial a_1}\varphi(||z_t - a_1||^2) & w_{22}\frac{\partial}{\partial a_2}\varphi(||z_t - a_2||^2) & \dots & w_{2p}\frac{\partial}{\partial a_p}\varphi(||z_t - a_p||^2) \\ \dots & \dots & \dots & \dots \\ w_{m1}\frac{\partial}{\partial a_1}\varphi(||z_t - a_1||^2) & w_{m2}\frac{\partial}{\partial a_2}\varphi(||z_t - a_2||^2) & \dots & w_{mp}\frac{\partial}{\partial a_1}\frac{\partial}{\partial a_p}\varphi(||z_t - a_p||^2) \end{pmatrix}.$$

Example 4.5.

Consider the classical problem of partitioning iris flowers into three classes, using four features [80]. The experimental data contain 50 instances of each class (a matrix 4×150), previously normalized according to the average and variance that were estimated for the entire set. The samples were randomly divided into two sets of training and testing with 25 elements of each class. The training set is denoted $L_3 = (X_1\ X_2\ X_3)$, where X_1, X_2, $X_3 \in R^{4\times 25}$ are matrices characterizing each class. It is required to develop an algorithm allowing determining to which class a particular pattern of Iris belongs. We use RBNN Eq. (4.174) with $m = 3, p = 4$.

Network training consisted of the evaluation of centers and the weights on the training set. We used two algorithms, namely, the GN method with a large parameter and the DTA in the iterative mode. Vectors β and α are interpreted as random variables. A priori information about the centers is determined from experimental data and on the output layer it is absent.

We put $E\beta = (\overline{v}_1^T, \overline{v}_2^T, \overline{v}_3^T)^T$, where $\overline{v}_1, \overline{v}_2, \overline{v}_3$ are the estimated average values of each class, which were estimated from samples X_1, X_2, X_3, respectively. For matrix $E((\beta - m_\beta)(\beta - m_\beta)^T)$ the estimate of features covariance matrix in all three classes derived from the sample is used, $L_3 = (X_1\ X_2\ X_3)$. The process of training and testing the network was repeated 100 times. In the result it was found that the GN method with the matrix of the intensities of the measurement noise 20.6 I_{75} diverged regardless of the value of μ in all 100 realizations. The DTA with the same noise intensity of measurements always converged, giving 89% of correct partition into classes with an average of 24 iterations.

CHAPTER 5

Diffuse Kalman Filter

Contents

5.1 PROBLEM STATEMENT

Let us consider a linear discrete system of the form

$$x_{t+1} = A_t x_t + B_t w_t, y_t = C_t x_t + D_t \xi_t, \quad t \in T = \{a, a+1, \ldots\}, \tag{5.1}$$

where $x_t \in R^n$ is the state vector, $y_t \in R^m$ is the measured output vector, $w_t \in R^r$ and $\xi_t \in R^l$ are the uncorrelated random processes with zero mean and the covariance matrices $E(w_t w_t^T) = I_r, E(\xi_t \xi_t^T) = I_l$, A_t, B_t, C_t, D_t are known matrices of appropriate dimensions.

Let the initial state of the system Eq. (5.1) x_a satisfies the following conditions:

A1. The random vector x_a is not correlated with w_t and ξ_t for $t \in T$.

A2. Without loss of generality it is assumed that the state vector elements are arranged so that there is a priori information regarding its $q \in \{1, \ldots, n-1\}$ first components at the initial moment $t = a$

$$m_i^a = E[x_{ia}], \quad \bar{s}_{ij}^a = E\left[(x_{ia} - m_i^a)(x_{ja} - m_j^a)\right], \quad i, j = 1, 2, \ldots, q. \tag{5.2}$$

A3. The random variables $x_{ia}, i = 1, 2, \ldots, q, q \geq 1$ are uncorrelated with $x_{ia}, i = q+1, q+2, \ldots, n$.

Diffuse Algorithms for Neural and Neuro-Fuzzy Networks.
DOI: http://dx.doi.org/10.1016/B978-0-12-812609-7.00005-6

A4. A priori information regarding the remaining $n-q$ components of x_a is absent and they are treated as random variables with zero mean and covariance matrix proportional to the large parameter $\mu > 0$, i.e.,

$$E(x_{ia}) = 0, \quad E(x_{ia}x_{ja}) = \mu \tilde{s}_{ij}^{a}, \quad i,j = q+1, q+2, \ldots, n, \tag{5.3}$$

where $\tilde{S}_a = ||\tilde{s}_{ij}^{a}||_{q+1}^{n} > 0$.

A5. If a priori information regarding the entire vector x_a is absent then the conditions A1 and A4 are satisfied with $q = 0$.

It is required to find the limit relations for the Kalman filter (KF) as $\mu \to \infty$ and to study their properties. We will call these relations the diffuse KF (the DKF).

Consider an alternative approach to the problem of the system state estimation in the absence of a priori information about the initial conditions. This approach assumes that the following conditions are satisfied.

B1. Without loss of generality it is assumed that the state vector elements are arranged so that there is a priori information Eq. (5.2) regarding its $q \in \{1, \ldots, n-1\}$ first components at the initial moment $t = a$.

B2. A priori information on the remaining $n-q$ components of x_a is absent and they can be either unknown constants or random variables statistical characteristics which are unknown.

B3. If x_{ia}, $i = 1, 2, \ldots, q$ are random variables, then they are uncorrelated with x_{ia}, $i = q+1, q+2, \ldots, n$, w_t and ξ_t for $t \in T$.

B4. If a priori information regarding the entire vector x_a is absent then the conditions B2 and B3 satisfy only with $q = 0$.

It is required to find conditions for the existence of the linear state estimate

$$\hat{x}_b = \Omega_{b-1,a} m_a + \sum_{t=a}^{b-1} Y_{b-1,t} y_t \tag{5.4}$$

from measurements $y_a, y_{a+1}, \ldots, y_{b-1}$, $b \in \{a+1, a+2, \ldots,\} \in T$ and a recursive representation for $\hat{x}_b$ calculation, where the matrices $Y_{b-1,t} \in R^{n \times m}$ and $\Omega_{b-1,a} \in R^{n \times q}$ do not depend on a priori information about the unknown vector x_a, $m^a = (m_1^a, \ldots, m_q^a)^T$. The estimate is found under the conditions that it is be unbiased

$$E(\hat{x}_b) = E(x_b) \tag{5.5}$$

and minimizes the criterion

$$E((x_b - \hat{x}_b)^T(x_b - \hat{x}_b)). \tag{5.6}$$

5.2 ESTIMATION WITH DIFFUSE INITIALIZATION

Consider the problem of estimating the state vector of the system Eq. (5.1) under the assumption that the conditions A1—A5 are fulfilled.

Under these assumptions, the linear unbiased state estimate of the system Eq. (5.1) with a minimum mean square error is determined from the equations system (the KF) [81]

$$\hat{x}_{t+1} = A_t\hat{x}_t + K_t(y_t - C_t\hat{x}_t), \quad \hat{x}_a = (m_a^T, 0_{1\times(n-q)})^T, \quad t\in T = \{a, a+1, \ldots\}, \tag{5.7}$$

where

$$K_t = A_tP_tC_t^TN_t^{-1}, \tag{5.8}$$

$$P_{t+1} = A_tP_tA_t^T - A_tP_tC_t^TN_t^{-1}C_tP_tA_t^T + V_{2t}, \quad P_a = block\ diag(\overline{S}_a, \mu\tilde{S}_a), \tag{5.9}$$

$$N_t = C_tP_tC_t^T + V_{1t}, \quad V_{1t} = D_tD_t^T, \quad V_{2t} = B_tB_t^T,$$

$$m_a = (m_1^a, \ldots, m_q^a)^T, \quad \overline{S}_a = ||\overline{s}_{ij}^a||_{q+1}^n \in R^{q\times q)}, \quad \tilde{S}_a = ||\tilde{s}_{ij}^a||_{q+1}^n \in R^{(n-q)\times(n-q)}.$$

We want to study the behavior of the KF for large values of μ and to find the limit relations as $\mu\to\infty$. We prove two auxiliary results which will be needed to solve this problem.

Lemma 5.1.

The estimation error covariance matrix P_t satisfies for $t\in T$ the relations

$$P_t = S_t + Q_t = S_t + R_tM_t^{-1}R_t^T, \tag{5.10}$$

where

$$S_{t+1} = A_tS_tA_t^T - A_tS_tC_t^TN_{1t}^{-1}C_tS_tA_t^T + V_{2t}, \quad S_a = block\ diag\,(\overline{S}_a, 0_{(n-q)\times(n-q)}), \tag{5.11}$$

$$R_{t+1} = A_{1t}R_t, \ A_{1t} = A_t - A_tS_tC_t^TN_{1t}^{-1}C_t, \ R_a = (e_{q+1}, \ldots, e_n), \tag{5.12}$$

$$M_{t+1} = M_t + R_t^TC_t^TN_{1t}^{-1}C_tR_t, \ M_a = \tilde{S}_a^{-1}/\mu, \tag{5.13}$$

$$Q_{t+1} = A_{1t}Q_tA_{1t}^T - A_{1t}Q_{2t}C_t^TN_t^{-1}C_tQ_tA_{1t}^T,\ Q_a = block\ diag\,(0_{q\times q}, \mu\overline{S}_a), \tag{5.14}$$

$$N_{1t} = C_tS_tC_t^T + V_{1t},\ N_t = C_t(S_t + Q_t)C_t^T + V_{1t},$$

$$S_t \in R^{n\times n},\ R_t \in R^{n\times(n-q)},\ M_t \in R^{(n-q)\times(n-q)},\ Q_t \in R^{n\times n},$$

Proof

The first expression in Eq. (5.10) follows from the known property [69] of the discrete Riccati matrix equation:

$$Q_t = \tilde{P}_t - \overline{P}_t, \quad t \in T,$$

where $\tilde{P}_t, \overline{P}_t$ are arbitrary solutions of Eq. (5.9). Let us prove the second expression. We first show that

$$Q_t = R_tM_t^{-1}R_t^T. \tag{5.15}$$

Substituting this expression into Eq. (5.14) gives

$$R_{t+1}M_{t+1}^{-1}R_{t+1}^T = A_tR_tM_t^{-1}R_t^TA_t^T - A_tR_tM_t^{-1}R_t^TC_t^TN_t^{-1}C_tR_tM_t^{-1}R_t^TA_t^T.$$

This relation is valid under the condition that M_t^{-1} satisfies the difference equation

$$M_{t+1}^{-1} = M_t^{-1} - M_t^{-1}R_t^TC_t^TN_t^{-1}C_tR_tM_t^{-1},\ M_a^{-1} = \tilde{S}_a\mu.$$

Transforming it using the identity Eq. (2.7) for $B = M_t^{-1}$, $C = C_tR_t$, $D = N_{1t}$, we obtain the expression

$$M_{t+1}^{-1} = (M_t + R_t^TC_t^TN_{1t}^{-1}C_tR_t)^{-1} \tag{5.16}$$

which implies Eqs. (5.13) and (5.10).

Lemma 5.2.

The KF gain matrix Eq. (5.8) satisfies the relation

$$K_t = A_t(S_t + Q_t)C_t^TN_t^{-1} = (A_tS_t + A_{1t}R_tM_{t+1}^{-1}R_t^T)C_t^TN_{1t}^{-1},\ t \in T. \tag{5.17}$$

Proof

The first expression in Eq. (5.17) follows from Lemma 5.1. Let us prove the second equality. We first show that

$$K_t = A_tP_tC_t^TN_t^{-1} = A_tS_tC_t^TN_{1t}^{-1} + A_{1t}Q_tC_t^TN_t^{-1}.$$

Since

$$P_t = S_t + Q_t$$

(Lemma 5.1) then it should be

$$A_t(S_t + Q_t)C_t^T = A_tS_tC_t^TN_{1t}^{-1}N_t + A_{1t}Q_tC_t^T.$$

Indeed, transforming the right side of this expression, we establish that

$$\begin{aligned} A_tS_tC_t^TN_{1t}^{-1}N_t + A_{1t}Q_tC_t^T &= A_tS_tC_t^T(I_m + N_{1t}^{-1}C_tQ_tC_t^T) \\ &\quad + (A_t - A_tS_tC_t^TN_{1t}^{-1}C_t)Q_tC_t^T \\ &= A_t(S_t + Q_t)C_t^T \end{aligned}$$

Using Eqs. (5.15) and (5.16) gives

$$\begin{aligned} A_{1t}Q_tC_t^TN_t^{-1} &= A_{1t}R_tM_t^{-1}R_t^TC_t^TN_t^{-1} \\ &= A_{1t}R_tM_{t+1}^{-1}(M_t + R_t^TC_t^TN_{1t}^{-1}C_tR_t)M_t^{-1}R_t^TC_t^TN_t^{-1} \\ &= A_{1t}R_tM_{t+1}^{-1}(R_t^TC_t^T + R_t^TC_t^TN_{1t}^{-1}C_tR_tM_t^{-1}R_t^TC_t^T)N_t^{-1} \\ &= A_{1t}R_tM_{t+1}^{-1}R_t^TC_t^TN_{1t}^{-1}(N_{1t} + C_tR_tM_t^{-1}R_t^TC_t^T)N_t^{-1} \\ &= A_{1t}R_tM_{t+1}^{-1}R_t^TC_t^TN_{1t}^{-1} \end{aligned}$$

This implies Eq. (5.17).

Theorem 5.1.

Let $\overline{T} = \{a, a+1, \ldots, b\}$ be an arbitrary bounded set. Then:

1. The following uniform asymptotic representations in $t \in \overline{T}$ are valid

$$P_t = R_t\tilde{S}_a(I_{n-q} - W_tW_t^+)R_t^T\mu + S_t + R_tW_t^+R_t^T + O(\mu^{-1}), \quad \mu \to \infty, \tag{5.18}$$

$$K_t = \begin{cases} 0 & C_t = 0 \\ K_t^{dif} + O(\mu^{-1}) & \mu \to \infty, \quad C_t \neq 0, \end{cases} \tag{5.19}$$

where

$$W_{t+1} = W_t + R_t^TC_t^TN_{1t}^{-1}C_tR_t, \quad W_a = 0_{(n-q)\times(n-q)}, \tag{5.20}$$

$$K_t^{dif} = (A_tS_t + A_{1t}R_tW_{t+1}^+R_t^T)C_t^TN_{1t}^{-1}, \tag{5.21}$$

S_t and R_t are the solutions of Eqs. (5.11) and (5.12), respectively.

2. For any $\varepsilon > 0$ the following condition is satisfied

$$P(||\hat{x}_t - \hat{x}_t^{dif}|| \geq \varepsilon) = O(\mu^{-1}), \ \ \mu \rightarrow \infty, \ \ t \in \overline{T}, \tag{5.22}$$

where

$$\hat{x}_{t+1}^{dif} = A_t \hat{x}_t^{dif} + K_t^{dif}(y_t - C_t \hat{x}_t^{dif}), \ \ \hat{x}_a^{dif} = (m_a^T, 0_{1 \times (n-q)})^T. \tag{5.23}$$

Proof

1. From Eq. (5.13) it follows that

$$M_t = \tilde{S}_a^{-1}/\mu + \sum_{s=a}^{t-1} R_t^T C_t^T N_{1t}^{-1} C_t R_t = \tilde{S}_a^{-1}/\mu + W_t.$$

Using Lemma 2.1 for $\Omega_t = M_t, \ \ \Omega_0 = \tilde{S}_a^{-1}, \ \ F_t = N_{1t}^{-1/2} C_t R_t$ we get

$$M_t^{-1} = \tilde{S}_a(I_{n-q} - W_t W_t^+)\mu + W_t^+ + O(\mu^{-1}), \ \ \mu \rightarrow \infty.$$

Substitution of this expression into Eq. (5.10) yields Eq. (5.18).

From Eq. (5.8) it follows that $K_t = 0$ if $C_t = 0$. Let $C_t \neq 0$. The use of Lemma 2.2 gives

$$(I_{n-q} - W_{t+1} W_{t+1}^+) R_t^T C_t^T N_{1t}^{-1/2} = 0$$

that together with the relation (5.17) implies Eq. (5.19).

2. Introducing the notations

$$e_t = \hat{x}_t - \hat{x}_t^{dif}, \ \ h_t = \hat{x}_t - x_t$$

and using Eqs. (5.1) and (5.7), we obtain

$$e_t = (A_t - K_t^{dif} C_t) e_{t-1} - (K_t^{dif} - K_t) C_t h_t + (K_t^{dif} - K_t) B_t w_t, \ \ e_0 = 0,$$

$$h_t = (A_t - K_t^{dif} C_t) h_{t-1} + K_t^{dif} B_t w_t, \ \ h_a = ((m^a - \overline{x}_a)^T, -\tilde{x}_a^T)^T,$$

where

$$\overline{x}_a = (x_{1a}, x_{2a}, \ldots, x_{qa})^T, \ \ \tilde{x}_a = (x_{q+1a}, x_{q+2a}, \ldots, x_{na})^T.$$

The matrix of second moments of the block vector $x_t = (e_t^T, h_t^T)^T$ satisfies the following matrix difference equation

$$Q_t = \tilde{A}_t Q_{t-1} \tilde{A}_t^T + L_t, \ \ Q_0 = block \ diag \ (0_{q \times q}, \overline{Q}_0),$$

where

$$\tilde{A}_t = \begin{pmatrix} A_t - K_t^{dif} C_t & -(K_t - K_t^{dif})C_t \\ 0 & A_t - K_t^{dif} C_t \end{pmatrix},$$

$$L_t = \begin{pmatrix} (K_t - K_t^{dif})R_t(K_t - K_t^{dif})^T & (K_t - K_t^{dif})R_t(K_t^{dif})^T \\ K_t^{dif} R_t(K_t - K_t^{dif})^T & K_t^{dif} R_t(K_t^{dif})^T \end{pmatrix},$$

$$R_t = B_t B_t^T, \ \overline{Q}_0 = E(h_a h_a^T).$$

Since

$$||K_s - K_t^{dif}|| = O(\mu^{-1}), \ t \in \overline{T}, \ \mu \to \infty$$

then this implies

$$E(e_t^T e_t) = O(\mu^{-1}), \ t \in \overline{T}, \ \mu \to \infty.$$

where $O(\mu^k)$ is a function such that $O(\mu^k)/\mu^k$ is bounded as $\mu \to \infty$. Using the Markov's inequality gives for any $\varepsilon > 0$

$$P(||e_t|| \geq \varepsilon) \leq E(||e_t||^2)/\varepsilon^2 = O(\mu^{-1}), \ \mu \to \infty, \ t \in \overline{T}.$$

The relations (5.20), (5.21), and (5.23) will be called the diffuse KF (DKF).

Consequence 5.1.
The diffuse component

$$P_t^{dif} = R_t \tilde{S}_a (I_{n-q} - W_t W_t^+) R_t^T \mu$$

vanishes when, $t \geq tr$, where $tr = \min_t\{t: W_t > 0, \ t = a, a+1, \ldots\}$.

Consequence 5.2.
The matrix K_t^{dif} does not depend on the diffuse component as opposed to the matrix P_t and as the function of μ is uniformly bounded in the norm for $t \in \overline{T}$ as $\mu \to \infty$.

Consequence 5.3.
Numerical implementation errors can result in the KF divergence for large values of μ. Indeed, let δW_t^+ be the error connected with calculations of the pseudoinverse W_t. Then by Theorem 5.1

$$K_t = R_t \tilde{S}_a (I_{n-q} - W_t(W_t^+ + \delta W_t^+)) R_t^T C_t^T N_{1t}^{-1} \mu + O(1), \ t \in \overline{T}, \ \mu \to \infty. \tag{5.24}$$

For $\delta W_t^+ \neq 0$ the matrix K_t becomes dependent on the diffuse component. Moreover, in this case K_t becomes proportional to a large parameter and so divergence is possible even if the continuity condition is satisfied and δW_t^+ is arbitrarily small in norm.

Let us consider the properties of the DKF optimality. We first prove the following auxiliary assertion.

Lemma 5.3.

There are the following representations

$$H_{t,s} = \Phi_{t,s} - R_t W_t^{-1} \sum_{i=s}^{t-1} R_i^T C_i^T N_{1i}^{-1} C_i \Phi_{i,s}, \quad t \geq t_{tr}, s < t, \tag{5.25}$$

$$(H_{s,t+1} \tilde{K}_t^{dif})^T = N_{1t}^{-1} C_t R_t W_s^{-1} R_s^T = f_{t,s}, \tag{5.26}$$

where $H_{t,s}$, $\Phi_{t,s}$, are determined from the equations systems

$$H_{t+1,s} = (A_t - K_t^{diff} C_t) H_{t,s} = \tilde{A}_t H_{t,s}, \quad H_{t,s} = I_n, \quad t \geq s, \tag{5.27}$$

$$\Phi_{t+1,s} = A_{1t} \Phi_{t,s}, \quad \Phi_{s,s} = I_n, \quad t > s. \tag{5.28}$$

Proof

Consider the auxiliary equations systems

$$G_{t,s} = (A_t - K_t^{dif} C_t)^T G_{t+1,s} = \tilde{A}_t^T G_{t+1,s}, \quad G_{s,s} = I_n, \tag{5.29}$$

$$Z_{t,s} = A_{1t}^T Z_{t+1,s} - C_t^T N_{1t}^{-1} C_t R_t W_s^{-1} R_s^T, \quad Z_{s,s} = I_n, \quad s \geq tr. \tag{5.30}$$

We at first show that

$$G_{t,s} = Z_{t,s}, \quad t = s-1, s-2, \ldots, \quad s \geq tr. \tag{5.31}$$

Iterating Eqs. (5.30) and (5.28) obtain

$$Z_{t,s} = A_{1t}^T A_{1t+1}^T \ldots A_{1s-1}^T - A_{1t}^T A_{1t+1}^T \ldots A_{1s-2}^T C_{s-1}^T f_{s-1,s}$$
$$- \ldots - C_t^T f_{t,s}, \; t \geq t_{tr}, \; t \geq s.$$

$$\Phi_{t,s} = A_{1t-1} A_{1t-2} \ldots A_{1s},$$

$$\Phi_{s,t}^T = (A_{1s-1} A_{1s-2} \ldots A_{1t})^T = A_{1t}^T A_{1t+1}^T \ldots A_{1s-1}^T.$$

In view of these expressions $Z_{t,s}$ will take the form

$$\begin{aligned} Z_{t,s} &= \Phi_{s,t}^T - \Phi_{s+1,t}^T C_{s-1}^T f_{s-1,s} - \cdots - C_t^T f_{t,s} \\ &= \Phi_{s,t}^T - \sum_{j=t}^{s-1} \Phi_{j,t}^T C_j^T N_{1j}^{-1} C_j R_j W_s^{-1} R_s^T, \quad t \geq t_{tr}, t \geq s. \end{aligned} \tag{5.32}$$

Represent Eq. (5.29) in the following equivalent form

$$G_{t,s} = (A_t - K_t^{dif} C_t)^T G_{t+1,s} = (\tilde{A}_{1t} - \tilde{K}_t^{dif} C_t)^T G_{t+1,s}, \quad G_{s,s} = I_n, \tag{5.33}$$

where

$$\tilde{K}_t^{dif} = R_{t+1} W_{t+1}^+ R_t^T C_t^T N_{1t}^{-1}.$$

From comparison of the right-hand side parts of Eqs. (5.30) and (5.33) it follows that Eq. (5.31) is performed if

$$(\tilde{K}_t^{dif})^T Z_{t+1,s} = N_t^{-1} C_t R_t W_s^{-1} R_s^T. \tag{5.34}$$

Substituting Eq. (5.32) into the left side of Eq. (5.34) obtain

$$\begin{aligned} (\tilde{K}_t^{dif})^T Z_{t+1,s} &= N_{1t}^{-1} C_t R_t W_{t+1}^+ R_{t+1}^T \left(\Phi_{s,t+1}^T - \sum_{j=t+1}^{s-1} \Phi_{j,t+1}^T C_j^T N_{1j}^{-1} C_j R_j W_s^{-1} R_s^T \right) \\ &= N_{1t}^{-1} C_t R_t W_{t+1}^+ \left(I_n - \sum_{j=t+1}^{s-1} R_j^T C_j^T N_{1j}^{-1} C_j R_j W_s^{-1} \right) R_s^T \\ &= N_{1t}^{-1} C_t R_t W_{t+1}^+ W_{t+1} W_s^{-1} R_s^T. \end{aligned} \tag{5.35}$$

In the derivation of Eq. (5.35) we used the identities

$$\begin{aligned} R_{t+1} &= \Phi_{t+1,a} R_a, \quad \Phi_{s,t+1} \Phi_{t+1,a} = \Phi_{s,a}, \\ \Phi_{j,t+1} R_{t+1} &= \Phi_{j,t+1} \Phi_{t+1,a} R_a = \Phi_{j,a} R_a = R_j. \end{aligned}$$

Let us transform Eq. (5.35) using the orthogonal decomposition $W_t = V_t V_t^T$, where

$$V_t = \left(R_a^T C_a^T N_a^{-1/2}, R_{a+1}^T C_{a+1}^T N_{a+1}^{-1/2}, \ldots, R_t^T C_t^T N_t^{-1/2} \right).$$

Let $l_{1t}, l_{2t}, \ldots, l_{k(t),t}$ be linearly independent columns of the matrix V_t. Using skeletal decomposition yields

$$V_t = L_t \Gamma_t,$$

where $L_t = (l_{1t}, l_{2t}, \ldots, l_{k(t),t}) \in R_{k(t)}^{(n-q)\times k(t)}$, $\Gamma_t \in R_{k(t)}^{k(t)\times m(t-a)}$, $R_r^{k\times l}$, are a set of $k \times l$ matrices of the rank r. Since $\tilde{\Gamma}_t = \Gamma_t \Gamma_t^T > 0$ is the Gram matrix constructed by linearly independent rows of the matrix Γ_t and $rank(L_t) = rank(\tilde{\Gamma}_t L_t^T)$ then

$$W_t = L_t \tilde{\Gamma}_t L_t^T, \quad W_t^+ = (L_t \tilde{\Gamma}_t L_t^T)^+ = (L_t^T)^+ \tilde{\Gamma}_t^{-1} L_t^+.$$

The validity of Eqs. (5.25) and (5.34) follows from the expressions

$$\begin{aligned} N_{1t}^{-1/2} C_t R_t W_t^+ W_t &= N_{1t}^{-1/2} C_t R_t (L_t^T)^+ \tilde{\Gamma}_t^{-1} L_t^+ L_t \tilde{\Gamma}_t L_t^T \\ &= N_{1t}^{-1/2} C_t R_t L_t (L_t^T L_t)^{-1} L_t^T = \Gamma_{tt}^T (L_t^T L_t)(L_t^T L_t)^{-1} L_t^T \\ &= N_{1t}^{-1/2} C_t R_t, \end{aligned}$$

$$G_{s,t} = \tilde{A}_s^T \tilde{A}_{s+1}^T \ldots \tilde{A}_{t-1}^T, \quad H_{t,s} = \tilde{A}_{t-1} \tilde{A}_{t-2} \ldots \tilde{A}_s, \quad H_{t,s} = G_{s,t}^T$$

and Eq. (5.26) from Eq. (5.34).

Theorem 5.2.

1. The estimate of the system state Eq. (5.1) obtained by the DKF is unbiased when $t \geq tr$.
2. There are the following relations

$$E(||x_t - \hat{x}_t||^2) \geq E(||x_t - \hat{x}_t^{dif}||^2), \tag{5.36}$$

$$P_t^{dif} = E((x_t - \hat{x}_t^{dif})(x_t - \hat{x}_t^{dif})^T) = S_t + R_t W_t^{-1} R_t^T, \quad t \geq t_{tr}, \tag{5.37}$$

where $\hat{x}_t$ is an unbiased estimate of the system state Eq. (5.1) when $t \geq t_{tr}$ defined by the equations system

$$\hat{x}_{t+1} = A_t \hat{x}_t + \tilde{K}_t(y_t - C_t \hat{x}_t), \quad \hat{x}_a = (m_a^T, 0_{1\times(n-q)})^T, \quad t \in T = \{a, a+1, \ldots\} \tag{5.38}$$

with arbitrary matrix gain $\tilde{K}_t \in R^{n\times m}$ that is independent of the unknown a priori information about initial conditions, P_t^{dif} is the covariance matrix of the estimation error.

Proof

1. Consider the system of equations

$$h_{t+1} = (A_t - K_t^{dif} C_t) h_t$$

with the initial condition $h_a = (0_{1\times q}, \tilde{x}^T)^T$, where $h_t = E(\hat{x}_t^{dif} - x_t)$, $\tilde{x} = (x_{a(q+1)}, x_{a(q+2)}, \ldots, x_{an})^T$ is an arbitrary vector.

Using Lemma 5.3, we establish that

$$h_t = H_{t,a}h_a = \left(\Phi_{t,a} - R_t W_t^{-1} \sum_{i=a}^{t-1} R_i^T C_i^T N_{1i}^{-1} C_i \Phi_{i,a}\right) R_a \tilde{x} = 0.$$

Whence it follows that the estimate $\hat{x}_t^{dif}$ is unbiased, i.e., $E(\hat{x}_t^{dif}) = E(x_t)$ when $t \geq t_{tr}$.

2. The covariance matrix of the estimation error

$$\tilde{P}_t = E((x_t - \hat{x}_t)(x_t - \hat{x}_t)^T)$$

satisfies the matrix equation

$$\tilde{P}_{t+1} = (A_t - \tilde{K}_t C_t)\tilde{P}_t(A_t - \tilde{K}_t C_t)^T + \tilde{K}_t V_{1t} \tilde{K}_t^T + V_{2t} \tag{5.39}$$

with some unknown initial matrix $\tilde{P}_{tr}$.

Complementing the right-hand side of Eq. (5.39) to the complete square, we get

$$\begin{aligned}\tilde{P}_{t+1} &= A_t \tilde{P}_t A_t^T - A_t \tilde{P}_t C_t^T N_t^{-1} C_t \tilde{P}_t A_t^T + V_{2t} \\ &\quad + (\tilde{K}_t - A_t \tilde{P}_t C_t^T N_t^{-1}) N_t (\tilde{K}_t - A_t \tilde{P}_t C_t^T N_t^{-1})^T.\end{aligned} \tag{5.40}$$

It follows from this that if we put $\tilde{K}_t = A_t \tilde{P}_t C_t^T N_t^{-1}$, then

$$\tilde{P}_{t+1} = A_t \tilde{P}_t A_t^T - A_t \tilde{P}_t C_t^T N_t^{-1} C_t \tilde{P}_t A_t^T + V_{2t} \tag{5.41}$$

and in this case $\tilde{P}_t \leq \overline{P}_t$, where $\overline{P}_t$ is arbitrary solution of the matrix Eq. (5.39).

From Lemma 5.1 it follows that the set of the equation solutions (5.41) includes the solution of the form

$$P_t^{dif} = S_t + Q_t = S_t + R_t W_t^{-1} R_t^T, \quad t > tr$$

with the initial condition $P_{tr}^{dif} = S_{tr} + R_{tr} W_{tr}^{-1} R_{tr}^T$. Thus $P_t^{dif} \leq \tilde{P}_t$.

Comment 5.2.1.

Comparing the KF with a large parameter μ described by expressions (5.7) and (5.9) with the DKF it is easy to see that from the point of view of computation that the DKF is significantly more expensive. Indeed, in the case of the DKF we need to iterate additionally two matrix Eqs. (5.11) and (5.12). In addition, there is a number of additional matrix operations related to the definition of the DKF gain. However, Consequence 5.3 shows that such complex structure of the DKF is compensated by its resistance to the accumulation of errors providing the estimates convergence.

5.3 ESTIMATION IN THE ABSENCE OR INCOMPLETE A PRIORI INFORMATION ABOUT INITIAL CONDITIONS

In this section we consider a problem of estimation of the state vector of the system Eq. (5.1) under the assumption that the conditions B1—B4 are fulfilled.

We show at first that the problem of unbiased estimation of the state vector Eq. (5.1) is due to a special control task. Based on this result, we find the existence conditions for an unbiased estimator and expressions describing its work.

Consider an auxiliary linear system of the form

$$Z_t = A_t^T Z_{t+1} - C_t^T U_t,\quad Z_b = I_n,\quad t = b-1, b-2, \ldots, a, \tag{5.42}$$

where $Z_t \in R^{n\times n}$, $U_t \in R^{m\times n}$ is a control.

Lemma 5.4.

Suppose that there is a control U_t bringing the system Eq. (5.42) in a state satisfying a condition

$$\Psi^T Z_a = 0,$$

where $\Psi = (e_{q+1}, \ldots, e_n)$, $e_i \in R^n$, e_i is the i-th unit vector, $i = 1, 2, \ldots, q$. Then, there is an unbiased state estimate of the system Eq. (5.1) of the form Eq. (5.4), where

$$Y_{b-1,t} = U_t^T,\quad \Omega_{b-1,a} = Z_a^T(e_1, \ldots, e_q). \tag{5.43}$$

Proof

Using the identity

$$x_b = Z_a^T x_a + \sum_{t=a}^{b-1}(Z_{t+1}^T x_{t+1} - Z_t^T x_t), \tag{5.44}$$

Eqs. (5.1) and (5.2) give

$$\begin{aligned} x_b &= Z_a^T x_a + \sum_{t=a}^{b-1}[Z_{t+1}^T(A_t x_t + B_t w_t) - (Z_{t+1}^T A_t - U_t^T C_t)x_t] \\ &= Z_a^T x_a + \sum_{t=a}^{b-1}(Z_{t+1}^T B_t w_t + U_t^T C_t x_t). \end{aligned} \tag{5.45}$$

Therefore, the expression for estimation error at the moment $t=b$ can be presented in the form

$$\begin{aligned} e_b = x_b - \hat{x}_b &= Z_a^T x_a - \Omega_{b-1,a} m_a + \sum_{t=a}^{b-1} (Z_{t+1}^T B_t w_t + U_t^T C_t x_t - Y_{b-1,t} y_t) \\ &= Z_a^T x_a - Z_a^T (e_1, \ldots, e_q) m_a + \sum_{t=a}^{b-1} (Z_{t+1}^T B_t w_t - U_t^T D_t \xi_t) \\ &= Z_a^T q_1 + Z_a^T q_2 + \sum_{t=a}^{b-1} (Z_{t+1}^T B_t w_t - U_t^T D_t \xi_t), \end{aligned} \tag{5.46}$$

where

$$q_1 = (x_a(1) - m_1^a, x_a(2) - m_2^a \ldots, x_a(q) - m_q^a, 0, 0, \ldots, 0)^T \in R^n,$$
$$q_2 = (0, 0, \ldots, 0, x_a(q+1), x_a(q+2), \ldots, x_a(n),)^T \in R^n.$$

Averaging the left- and right-hand sides of this expression and using the condition $\Psi^T Z_a = 0$ we get

$$\begin{aligned} E(e_b) &= E(Z_a^T q_1) + E(Z_a^T q_2) = Z_a^T E(q_2) \\ &= Z_a^T \Psi \{E[x_a(q+1)],\ E[x_a(q+2)], \ldots, E[x_a(n)]\}^T = 0. \end{aligned}$$

Lemma 5.5.
Let

$$W_{a,b} = \sum_{j=a}^{b-1} \Psi^T \Phi_{j,a}^T C_j^T C_j \Phi_{j,a} \Psi > 0, \tag{5.47}$$

where the matrix $\Phi_{t,a}$ is determined by the system

$$\Phi_{t+1,a} = A_t \Phi_{t,a}, \quad \Phi_{a,a} = I_n, \quad t \in T.$$

Then:

1. There is an unbiased estimate of the state vector of the system Eq. (5.1) at the moment $t=b$.
2. If additionally for any nonzero vector $p \in R^{n-q}$

$$A_{b-1} \ldots A_a (0_{1 \times q}, p^T)^T \neq 0 \tag{5.48}$$

then the condition Eq. (5.47) is the necessary and sufficient condition for the existence of an unbiased estimator of the system state Eq. (5.1) at the moment $t=b$.

Proof

1. We show that when performing Eq. (5.47) there exists a control referred to Lemma 5.4. Iterating Eq. (5.42) obtain

$$Z_t = G_{t,b} Z_b - \sum_{j=t}^{b-1} G_{t,j} C_j^T U_j, \quad t = b-1, b-2, \ldots, a, \tag{5.49}$$

where the matrix $G_{t,s}$ is determined by the system

$$G_{t,s} = A_t^T G_{t+1,s}, \quad G_{s,s} = I_n, \quad s \geq t.$$

Using the boundary condition $\Psi^T Z_a = 0$, we find with the help of Eq. (5.49)

$$0 = \Psi^T G_{a,b} Z_b - \sum_{j=a}^{b-1} \Psi^T G_{a,j} C_j^T U_j. \tag{5.50}$$

We seek a solution of this system in the form

$$U_t = C_t G_{a,t}^T \Psi L,$$

where $L \in R^{(n-q) \times n}$ is a constant, unknown matrix. It follows from Eq. (5.50) that

$$L = \tilde{W}_{a,b}^{-1} \Psi^T G_{a,b} Z_b$$

provided that $\tilde{W}_{a,b} > 0$, where

$$\tilde{W}_{a,b} = \sum_{j=a}^{b-1} \Psi^T G_{a,j} C_j^T C_j G_{a,j}^T \Psi.$$

But as

$$G_{t,b} = A_t^T A_{t+1}^T \ldots A_{b-2}^T A_{b-1}^T, \quad \Phi_{b,t} = A_{b-1} A_{b-2} \ldots A_{t-1} A_t$$

then $G_{t,b}^T = \Phi_{b,t}$, $W_{a,b} = \tilde{W}_{a,b}$ and the control is given by the expression

$$U_t = C_t G_{a,t}^T \Psi \tilde{W}_{a,b}^{-1} \Psi^T G_{a,b} = C_t \Phi_{t,a} \Psi W_{a,b}^{-1} \Psi^T \Phi_{b,a}^T. \tag{5.51}$$

The statement follows now from Lemma 5.4.

2. Suppose that there are $Y_{b-1,t}$, $\Omega_{b-1,a}$ such that the estimate Eq. (5.4) is an unbiased and performed Eq. (5.48), but $W_{a,b}$ is singular. Then there is a vector $p \neq 0$, $p \in R^{n-q}$, such that $W_{a,b}p = 0$. Let us denote $\varsigma_t = C_t \Phi_{t,a} \Psi p$. As

$$\sum_{t=a}^{b-1} \varsigma_t^T \varsigma_t = p^T W_{a,b} p$$

then $\varsigma_t = 0$ for any $t \in \{a, \ldots, b-1\}$. Let $x_a = (\eta^T, p^T)^T$, where $\eta \in R^q$ is any random vector with zero mean. Using Eqs. (5.1), (5.43) gives the relation

$$\begin{aligned} E(\hat{x}_b) &= \Omega_{b-1,a} E(\eta) + E\left(\sum_{t=a}^{b-1} Y_{b-1,t} y_t \right) \\ &= \sum_{t=a}^{b-1} Y_{b-1,t} E(y_t) = \sum_{t=a}^{b-1} U_t^T C_t \Phi_{t,a} E(x_a) \\ &= \sum_{t=a}^{b-1} U_t^T C_t \Phi_{t,a} \Psi p = \sum_{t=a}^{b-1} U_t^T \varsigma_t = 0 \end{aligned}$$

that is impossible if the condition Eq. (5.48) holds since

$$E(x_b) = \Phi_{b,a}(0_{1 \times q}, p^T)^T = A_{b-1} A_{b-2} \ldots A_a (0_{1 \times q}, p^T)^T \neq 0.$$

Note that if the matrix A_t, $t \in T$ is not singular (as an example, the system Eq. (5.1) is obtained as a result of a continuous system sampling), then Eq. (5.48) is performed automatically.

An unbiased estimate of the system state Eq. (5.1) is determined by the expression

$$\hat{x}_b = Z_a^T (m_a^T, 0_{1 \times (n-q)})^T + \sum_{t=a}^{b-1} \Phi_{b,a} \Psi W_{a,b-1}^{-1} \Psi^T \Phi_{t,a}^T C_t^T y_t,$$

where

$$Z_a = \Phi_{b,a}^T - \sum_{j=a}^{b-1} \Phi_{j,a}^T C_j^T C_j \Phi_{t,a} \Psi W_{a,b}^{-1} \Psi^T \Phi_{b,a}^T.$$

This assertion follows from Lemma 5.4 and the expressions (5.49) and (5.51).

We will show that under the made assumptions the determining problem of the optimal estimate of the system state Eq. (5.1) (unbiased estimate minimizing the criterion Eq. (5.6)) is a special dual optimal control problem. Based on this result, it has found recurrent presentation for evaluating of the system state Eq. (5.1).

Let us consider the system Eq. (5.42) together with the criterion of quality

$$J(U_t) = tr\left[Z_a^T \tilde{S}_a Z_a^T + \sum_{t=a}^{b-1}(Z_{t+1}^T V_{2t} Z_{t+1} + U_t^T V_{1t} U_t)\right], \tag{5.52}$$

where $S_a = blockdiag(\overline{S}_a, 0_{n-q,n-q})$. It is required to choose a control U_t from the condition of minimum of the criterion Eq. (5.52) along the system solutions Eq. (5.42).

From Lemma 5.4 it follows that if the following conditions $\Psi^T Z_a = 0$, $Y_{b-1,t} = U_t^T$, $\Omega_{b-1,a} = Z_a^T(e_1, \ldots, e_q)$ are held then the state estimate of the system Eq. (5.1) at the time $t = b$ is unbiased and the use of Eq. (5.46) gives

$$\begin{aligned} x_b - \hat{x}_b &= Z_a^T(x_{1a} - m_1^a, x_{2a} - m_2^a, \ldots, x_{qa} - m_q^a, 0, \ldots, 0)^T \\ &\quad + \sum_{t=a}^{b-1}(Z_{t+1}^T B_t w_t - U_t^T D_t \xi_t). \end{aligned}$$

Whence it follows that

$$\begin{aligned} E[(x_b - \hat{x}_b)(x_b - \hat{x}_b)^T] &= Z_a^T \tilde{S}_a Z_a^T + \sum_{t=a}^{b-1}(Z_{t+1}^T V_{2t} Z_{t+1} + U_t^T V_{1t} U_t), \\ J(U_t) &= E[(x_b - \hat{x}_b)^T(x_b - \hat{x}_b)]. \end{aligned} \tag{5.53}$$

Thus, the unbiased state estimate of Eq. (5.1), minimizing performance criterion Eq. (5.6), can be found from the solution of the formulated control problem with the matrices appearing in the description of Eq. (5.4) $Y_{b-1,t}$, $\Omega_{b-1,a}$.

We at first prove some auxiliary statements.

Lemma 5.6.

Let $W_{a,b} = \sum_{j=a}^{b-1} \Psi^T \Phi_{j,a}^T C_j^T C_j \Phi_{j,a} \Psi > 0$. Then, the optimal program control and the minimum value of the quality criterion in the problems (5.42) and (5.52) are determined by expressions

$$U_t^p = N_t^{-1} C_t(S_t A_t^T Z_{t+1} + R_t M_{a,b}^{-1} R_b^T), \tag{5.54}$$

$$J_{\min} = Tr(S_b + R_b M_{a,b}^{-1} R_b^T), \tag{5.55}$$

where

$$M_{a,b} = \sum_{t=a}^{b-1} R_t^T C_t^T N_t^{-1} C_t R_t, \tag{5.56}$$

$$Z_t = \Lambda_{b,t}^T - \sum_{j=t}^{b-1} \Lambda_{j,t}^T C_j^T N_j^{-1} C_j R_j M_{a,b}^{-1} R_b^T, \tag{5.57}$$

$$\Lambda_{t+1,s} = A_{1t}\Lambda_{t,s}, \quad \Lambda_{s,s} = I_n, \quad t \geq s, \tag{5.58}$$

$$R_{t+1} = A_{1t} R_t, \quad R_a = \Psi, \tag{5.59}$$

$$S_{t+1} = A_t S_t A_t^T - A_t S_t C_t^T N_t^{-1} C_t S_t A_t^T + V_{2t}, \quad S_a = \overline{S}_a, \tag{5.60}$$

$$A_{1t} = A_t - A_t S_t C_t^T N_t^{-1} C_t, \quad N_t = C_t S_t C_t^T + V_{1t}. \tag{5.61}$$

Proof

Consider the identity

$$I = -\sum_{t=a}^{b-1} (Z_{t+1}^T S_{t+1} Z_{t+1} - Z_t^T S_t Z_t) + Z_b^T S_b Z_b - Z_a^T S_a Z_a = 0.$$

Transforming this with the help of Eqs. (5.42) and (5.60), we obtain

$$\begin{aligned} I = &-\sum_{t=a}^{b-1} \big[Z_{t+1}^T (A_t S_t A_t^T - A_t S_t C_t^T N_t^{-1} C_t S_t A_t^T + V_{2t}) Z_{t+1} \\ &- (A_t^T Z_{t+1} - C_t^T U_t)^T S_t (A_t^T Z_{t+1} - C_t^T U_t) \big] + Z_b^T S_b Z_b - Z_a^T S_a Z_a \\ = &-\sum_{t=a}^{b-1} \big[Z_{t+1}^T (-A_t S_t C_t^T N_t^{-1} C_t S_t A_t^T + V_{2t}) Z_{t+1} - Z_{t+1} A_t S_t C_t^T U_t \\ &+ U_t^T C_t S_t A_t^T Z_{t+1} - U_t^T C_t S_t C_t^T U_t \big] + Z_b^T S_b Z_b - Z_a^T S_a Z_a = 0. \end{aligned}$$

Whence it follows that

$$\begin{aligned} J(U_t) &= tr \left[Z_a^T \tilde{S}_a Z_a^T + \sum_{t=a}^{b-1} (Z_{t+1}^T V_{2t} Z_{t+1} + U_t^T V_{1t} U_t) + I \right] \\ &= tr \left[Z_b^T S_b Z_b + \sum_{t=a}^{b-1} (U_t - N_t^{-1} C_t S_t A_t^T Z_{t+1})^T N_t (U_t - N_t^{-1} C_t S_t A_t^T Z_{t+1}) \right]. \end{aligned} \tag{5.62}$$

We will search a control in the form

$$U_t = N_t^{-1} C_t S_t A_t^T Z_{t+1} + N_t^{-1/2} \tilde{U}_t. \tag{5.63}$$

Substituting this expression in Eqs. (5.42) and (5.62) gives

$$Z_t = (A_t^T - C_t^T N_t^{-1} C_t S_t A_t^T) Z_{t+1} - C_t^T N_t^{-1/2} \tilde{U}_t = A_{1t}^T Z_{t+1} - C_t^T N_t^{-1/2} \tilde{U}_t, \tag{5.64}$$

$$Z_b = I_n, \ \Psi^T Z_a = 0, \quad t = b-1, b-2, \ldots, a, \tag{5.65}$$

$$J(\tilde{U}_t) = Tr\left\{ \sum_{t=a}^{b-1} \tilde{U}_t^T \tilde{U}_t + S_b \right\}. \tag{5.66}$$

Let us find a program control which moves the system Eq. (5.64) in such state when

$$\Psi^T Z_a = 0.$$

We have

$$Z_t = G_{t,b} Z_b - \sum_{j=t}^{b-1} G_{t,j} C_j^T N_t^{-1/2} \tilde{U}_j, \quad t = b-1, b-2, \ldots, a, \tag{5.67}$$

where

$$G_{t,s} = A_{1t}^T G_{t+1,s}, \quad G_{s,s} = I_n, \quad s \geq t.$$

From this, using the boundary condition $\Psi^T Z_a = 0$, we get

$$0 = \Psi^T G_{a,b} Z_b - \sum_{j=a}^{b-1} \Psi^T G_{a,j} C_j^T N_j^{-1/2} U_j. \tag{5.68}$$

We seek a solution Eq. (5.68) in the form

$$\tilde{U}_t = N_t^{-1/2} C_t G_{a,t}^T \Psi L,$$

where $L \in R^{(n-q) \times n}$ is a constant, unknown matrix. Substituting this expression in Eq. (5.68), we obtain

$$L = \left(\sum_{j=a}^{b-1} \Psi^T G_{a,j} C_j^T N_t^{-1/2} C_j G_{a,j}^T \Psi \right)^{-1} \Psi^T G_{a,b} Z_b$$

$$= \left(\sum_{j=a}^{b-1} \Psi^T \Lambda_{j,a}^T C_j^T N_j^{-1/2} C_j \Lambda_{j,a} \Psi \right)^{-1} \Psi^T \Lambda_{b,a}^T Z_b = M_{a,b}^{-1} \Psi^T \Lambda_{a,b}^T Z_b,$$

if $M_{a,b} > 0$. From whence it follows

$$\tilde{U}_t = N_t^{-1/2} C_t R_t M_{a,b}^{-1} R_b^T Z_b \tag{5.69}$$

and the representation Eq. (5.54) for U_t.

We show that $\tilde{U}_t$ really minimizes the criterion Eq. (5.53) on the system solutions of Eq. (5.42). Suppose that $\hat{U}_t$ is arbitrary control such that $\Psi^T Z_a = 0$. From Eq. (5.68) it follows that

$$\sum_{j=a}^{b-1} \Psi^T \Lambda_{j,t}^T C_j^T N_j^{-1/2} (\tilde{U}_j - \hat{U}_j) = 0.$$

Multiplying this expression left by the matrix $\Lambda_{a,b}^T \Psi M_{a,b-1}^{-1}$, we obtain the equality

$$\sum_{j=a}^{b-1} \tilde{U}_j^T (\tilde{U}_j - \hat{U}_j) = 0.$$

It follows from this

$$Tr\left[\sum_{j=t}^{b-1} (\tilde{U}_j - \hat{U}_j)^T (\tilde{U}_j - \hat{U}_j)\right] = Tr\left[\sum_{j=t}^{b-1} (\tilde{U}_j^T \tilde{U}_j - 2\tilde{U}_j^T \frown U_j + \hat{U}_j^T \hat{U}_j)\right]$$

$$= Tr\left(\sum_{j=t}^{b-1} \hat{U}_j^T \hat{U}_j\right) - Tr\left(\sum_{j=t}^{b-1} \tilde{U}_j^T \tilde{U}_j\right) \geq 0.$$

The expression (5.55) for the minimum value of the quality criterion follows from Eqs. (5.66) and (5.69).

Lemma 5.7.
The condition

$$W_{a,b} = \sum_{j=a}^{b-1} \Psi^T \Phi_{j,a}^T C_j^T C_j \Phi_{j,a} \Psi > 0$$

is sufficient for the existence of the optimal control. If for any nonzero vector $p \in R^{n-q}$

$$A_{b-1} A_{b-2} \ldots A_a (0_{1 \times q}, p^T)^T \neq 0,$$

then it is necessary and sufficient.

Proof

Sufficiency. Let us show that from $W_{a,b} > 0$ it follows $M_{a,b} > 0$. Let $W_{a,b} > 0$ but $M_{a,b}$ is singular. In this case there is a nonzero vector $p \in R^{n-q}$ such that $M_{a,b}\, p = 0$ and at the same time $C_t R_t p = 0,\ \ t \in \{a, \ldots, b-1\}$. We show that from this will follow the equalities

$$C_t R_t p = C_t A_{t-1} \ldots A_a \Psi p = C_t \Phi_{t,a} \Psi p = 0, \quad t \in \{a, \ldots, b-1\}. \tag{5.70}$$

As

$$C_a R_a p = C_a \Psi p = 0,$$

$$C_{a+1} R_{a+1} p = C_{a+1} A_{1a} \Psi p = C_{a+1}(A_a - A_a S_a C_a^T N_a^{-1} C_a) \Psi p = C_{a+1} A_a \Psi p = 0$$

then at $t = a$ and $t = a+1$ they are carried out. Suppose that these relations are valid at $a+2, \ldots, t$ and prove then that $C_t R_t p = 0$ at $t+1$.

We have

$$\begin{aligned} C_{t+1} R_{t+1} p &= C_{t+1}(A_t - A_t S_t C_t^T N_t^{-1} C_t) R_t p \\ &= C_{t+1} A_t R_t p = C_{t+1} A_t (A_{t-1} - A_{t-1} S_{t-1} C_{t-1}^T N_{t-1}^{-1} C_{t-1}) R_{t-1} p \\ &= C_{t+1} A_t A_{t-1} R_{t-1} p \\ &= C_{t+1} \Phi_{t+1,a} \Psi p = 0 \end{aligned}$$

that is impossible since

$$p^T W_{a,b} p = \sum_{j=a}^{b-1} p^T \Psi^T \Phi_{j,a}^T C_j^T C_j \Phi_{j,a} \Psi p > 0$$

and therefore it should be $C_t \Phi_{t,a} \Psi p > 0$ if $t \in \{a, \ldots, b-1\}$.

Necessity. Let the optimal control U_t exist but the matrix $W_{a,b}$ is singular. Then there is a nonzero vector $p \in R^{n-q}$ such that $W_{a,b} p = 0$ and at the same

$$C_t \Phi_{t,a} \Psi p = 0, \quad t \in \{a, \ldots, b-1\}.$$

Using the boundary condition $\Psi^T Z_a = 0$ and the representation

$$Z_t = \Phi_{b,t}^T - \sum_{j=t}^{b-1} \Phi_{j,t}^T C_j^T U_j, \quad t = b-1, b-2, \ldots, a$$

for the solutions of Eq. (5.42), we find

$$p^T \Psi^T \Phi_{b,a}^T = \sum_{j=a}^{b-1} p^T \Psi^T \Phi_{j,a}^T C_j^T U_j = 0,$$

that is impossible since

$$\Phi_{b,a}\Psi p = A_{b-1}\ldots A_a(0_q^T, p^T)^T \neq 0.$$

Let us denote: U_t^p is the optimal program control, U_t^f is the optimal feedback control.

Lemma 5.8.
Let $W_{a,b} > 0$. Then the optimal feedback control is determined by the expression

$$U_t^f = N_t^{-1}C_t(S_tA_t^T + R_tM_{a,t+1}^+R_{t+1}^T)Z_{t+1}, \quad t = b-1, b-2, \ldots, a. \quad (5.71)$$

Proof
Replace in Eq. (5.54) b by $t+1$, Z_b, by Z_{t+1} and $M_{a,b}^{-1}$ by $M_{a,t+1}^+$ to obtain the feedback control which is determined by Eq. (5.71). We show now that this control indeed solves our problem. Consider the system Eq. (5.42) under the action of the controls Eqs. (5.54) and (5.71).

We have

$$Z_t^p = A_t^TZ_{t+1}^p - C_t^TU_t^p, \quad (5.72)$$

$$Z_t^f = (A_t^T - C_t^TG_t)Z_{t+1}^f, \quad (5.73)$$

where

$$G_t = N_t^{-1}C_t(S_tA_t^T + R_tM_{a,t+1}^+R_{t+1}^T) = G_{1t} + G_{2t}.$$

Let us show that Z_t^p satisfies Eq. (5.73). Substituting U_t^p in Eq. (5.72) gives

$$Z_t^p = A_{1t}^TZ_{t+1}^p - C_t^TN_t^{-1}C_tR_tM_{a,b}^{-1}R_b^T.$$

From whence we find

$$Z_t^p = \Lambda_{b,t}^T - \sum_{j=t}^{b-1}\Lambda_{j,t}^TC_j^TN_j^{-1}C_jR_jM_{a,b}^{-1}R_b^T.$$

We have

$$\begin{aligned}
G_{2t}Z_{t+1}^p &= N_t^{-1}C_tR_tM_{a,t+1}^+R_{t+1}^T\left(\Lambda_{b,t+1}^T - \sum_{j=t+1}^{b-1}\Lambda_{j,t+1}^TC_j^TN_j^{-1}C_jR_jM_{a,b}^{-1}R_b^T\right) \\
&= N_t^{-1}C_tR_tM_{a,t+1}^+\left(I_n - \sum_{j=t+1}^{b-1}R_j^TC_j^TN_j^{-1}C_jR_jM_{a,b}^{-1}\right)R_b^T. \\
&= N_t^{-1}C_tR_tM_{a,t+1}^+M_{a,t+1}M_{a,b}^{-1}R_b^T.
\end{aligned} \quad (5.74)$$

The expression (5.74) follows from the identities

$$R_{t+1} = \Lambda_{t+1,a}\Psi, \quad \Lambda_{b,t+1}\Lambda_{t+1,a} = \Lambda_{b,a},$$

$$\Lambda_{j,t+1}R_{t+1} = \Lambda_{j,t+1}\Lambda_{t+1,a}\Psi = \Lambda_{j,a}\Psi = R_j.$$

Let us transform Eq. (5.74) using the orthogonal decomposition $M_{a,t} = V_t V_t^T$, where

$$V_t = (R_a^T C_a^T N_a^{-1/2}, \; R_{a+1}^T C_{a+1}^T N_{a+1}^{-1/2}, \ldots, R_t^T C_t^T N_t^{-1/2}). \tag{5.75}$$

Let $l_{1t}, l_{2t}, \ldots, l_{k(t),t}$ be linearly independent columns of the matrix V_t. Using skeletal decomposition yields

$$V_t = L_t \Gamma_t,$$

where $L_t = (l_{1t}, l_{2t}, \ldots, l_{k(t),t}) \in R_{k(t)}^{(n-q)\times k(t)}$, $\Gamma_t \in R_{k(t)}^{k(t)\times m(t-a)}$, $R_r^{k\times l}$ are a set of $k \times l$ matrices of the rank r. Since $\tilde{\Gamma}_t = \Gamma_t\Gamma_t^T > 0$ is the Gram matrix constructed by linearly independent rows of the matrix Γ_t and $rank(L_t) = rank(\tilde{\Gamma}_t L_t^T)$ then

$$M_{a,t} = L_t\tilde{\Gamma}_t L_t^T, \quad M_{a,t}^+ = (L_t\tilde{\Gamma}_t L_t^T)^+ = (L_t^T)^+\tilde{\Gamma}_t^{-1}L_t^+.$$

Substituting these expressions in Eq. (5.74) and using the representation $L_t^+ = (L_t^T L_t)^{-1}L_t^T$, we find

$$\begin{aligned} G_{2t}Z_{t+1}^p &= N_t^{-1}C_tR_t(L_t^T)^+\tilde{\Gamma}_t^{-1}L_t^+L_t\tilde{\Gamma}_tL_t^TM_{a,b}^{-1}R_b^T \\ &= N_t^{-1}C_tR_tL_t(L_t^TL_t)^{-1}L_t^TR_t^TM_{a,b}^{-1}R_b^T. \end{aligned} \tag{5.76}$$

From Eq. (5.75) it follows that $R_t^T C_t^T \tilde{N}_t = L_t\Gamma_{1t}$, where Γ_{1t} is some rectangular matrix. Substituting this expression into Eq. (5.76) gives

$$G_{2t}Z_{t+1}^p = \tilde{N}_t\Gamma_1^T L_t^T L_t(L_t^T L_t)^{-1}L_t^T R_t^T M_{a,b}^{-1}R_b^T = N_t^{-1}C_tR_tM_{a,b}^{-1}R_b^T.$$

It follows from this the assertion.

Theorem 5.3.

1. The state of the system

$$\hat{x}_{t+1} = A_t\hat{x}_t + K_t(y_t - C_t\hat{x}_t), \quad \hat{x}_a = (\tilde{m}_a^T, 0_{n-q}^T)^T, \quad t \in T \tag{5.77}$$

for $t \geq t_{tr}$, $t_{tr} = \min_t\{t: W_{a,t} > 0, \;\; t = a, a+1, \ldots\}$ is the optimal estimate, where

$$K_t = (A_tS_t + A_{1t}R_tM_{a,t+1}^+R_t^T)C_t^TN_t^{-1}, \tag{5.78}$$

$$S_{t+1} = A_t S_t A_t^T - A_t S_t C_t^T N_t^{-1} C_t S_t A_t^T + V_{2t}, \quad S_a = \overline{S}_a, \tag{5.79}$$

$$R_{t+1} = (A_t - A_t S_t C_t^T N_t^{-1} C_t) R_t, \quad R_a = (e^{q+1}, \ldots, e^n), \tag{5.80}$$

$$M_{a,t+1} = M_{a,t} + R_t^T C_t^T N_t^{-1} C_t R_t, \quad M_{a,a} = 0. \tag{5.81}$$

2. The estimation error covariance matrix for $t \geq t_{tr}$ is given by the expression

$$P_t = S_t + R_t M_{a,t}^{-1} R_t^T. \tag{5.82}$$

Proof

1. Put in Eq. (5.4)

$$Y_{b-1,t} = (U_t^p)^T, \quad \Omega_{b-1,a} = Z_a^T(e^1, \ldots, e^q), \quad b \geq t_{tr},$$

where U_t^p and Z_a were determined earlier in Lemma 5.6. This implies the optimality of the estimate at the moment $t = b$ and the relation

$$\hat{x}_b = \sum_{t=a}^{b-1} (U_t^p)^T y_t + Z_a^T(\tilde{m}_a^T, 0_{n-q}^T)^T.$$

From Lemma 5.7 it follows that

$$U_t^p = U_t^f = N_t^{-1} C_t (S_t A_t^T + R_t M_{a,t+1}^+ R_{t+1}^T) Z_{t+1}, \quad t = b-1, b-2, \ldots, a.$$

In view of this

$$\hat{x}_t = \sum_{s=a}^{t-1} (U_s^f)^T y_s + Z_a^T(\tilde{m}_a^T, 0_{n-q}^T)^T = \sum_{s=a}^{t-1} H_{s+1,t}^T K_s y_s + H_{a,t}^T(\tilde{m}_a^T, 0_{n-q}^T)^T, \tag{5.83}$$

where

$$H_{s,t} = (A_s^T - C_s^T K_s^T) H_{s+1,t}, \quad H_{t,t} = I_n, \quad s \leq t, \quad t \in T.$$

By Eq. (5.83) we find

$$\hat{x}_{t+1} - \hat{x}_t = K_t y_t + \sum_{s=a}^{t} (H_{s+1,t+1}^T - H_{s+1,t}^T) K_s y_s + (H_{a,t+1}^T - H_{a,t}^T)(\tilde{m}_a^T, 0_{n-q}^T)^T. \tag{5.84}$$

Using the identity

$$H_{s+1,t+1} = H_{s+1,t}H_{t,t+1},\quad H_{a,t+1} = H_{a,t}H_{t,t+1},\quad H_{t,t+1} = A_t^T - C_t^T K_t^T,$$

we transform this expression to the form which coincides with Eq. (5.77)

$$\begin{aligned}\hat{x}_{t+1} - \hat{x}_t &= K_t y_t + (H_{t,t+1}^T - I_n)\sum_{s=a}^{t-1} H_{s+1,t}^T K_s y_s + (H_{t,t+1}^T - I_n) H_{a,t}^T (\tilde{m}_a^T, 0_{n-q}^T)^T \\ &= K_t y_t + (A_t - K_t C_t - I_n)\left[\sum_{s=a}^{t-1} H_{s+1,t}^T K_s y_s + H_{a,t}^T (\tilde{m}_a^T, 0_{n-q}^T)^T\right] \\ &= K_t y_t + (A_t - K_t C_t - I_n)\hat{x}_t.\end{aligned}$$

Putting in Eq. (5.83) $t = a$ gives $\hat{x}_a = H_{a,a}^T (\tilde{m}_a^T, 0_{n-q}^T)^T = (\tilde{m}_a^T, 0_{n-q}^T)^T$.

2. It follows from the expression $Tr(S_t + R_t M_{a,t}^{-1} R_t^T) = E((x_t - \hat{x}_t)^T (x_t - \hat{x}_t))$.

Comment 5.3.1.

We have established the following methodically important result. The diffuse initialization of the KF leads to identical relations obtained by solving a special optimization problem under the assumption that a priori information about the initial state of the system Eq. (5.1) is incomplete or absent.

5.4 SYSTEMS STATE RECOVERY IN A FINITE NUMBER OF STEPS

Consider the system Eq. (5.1) in the absence of random disturbances

$$x_{t+1} = A_t x_t,\quad y_t = C_t x_t,\quad t \in T = \{a, a+1, \ldots\}. \tag{5.85}$$

It is required to restore the vector x_t, using the observations y_t, $t \in T$. To solve this problem we use the observer based on the design of the proposed the DKF

$$\hat{x}_{t+1}^{dif} = A_t \hat{x}_t^{dif} + K_t^{dif}(y_t - C_t \hat{x}_t^{dif}),\ \hat{x}_a = 0,\quad t \in T = \{a, a+1, \ldots\}, \tag{5.86}$$

where

$$\begin{aligned}K_t^{dif} &= A_t \Phi_{t,a} W_{t+1}^{+} \Phi_{t,a}^T C_t^T, \\ W_{t+1} &= W_t + \Phi_{t,a}^T C_t^T C_t \Phi_{t,a},\ W_a = 0_{(n-q)\times(n-q)}.\end{aligned}$$

This observer will restore the system state in a finite number of steps. Indeed, the recovery error

$$h_t = \hat{x}_t - x_t$$

satisfies the equations system

$$h_{t+1} = (A_t - K_t^{dif} C_t)h_t$$

with arbitrary unknown initial conditions. Using Lemma 5.3 with $S_t = 0,\ V_{1t} = 0_{m \times m}$ gives

$$H_{t,s} = \Phi_{t,s} - \Phi_{t,s} W_t^{-1} \sum_{i=s}^{t-1} \Phi_{i,s}^T C_i^T C_i \Phi_{i,s},$$

where

$$\Phi_{t+1,s} = A_t \Phi_{t,s},\quad \Phi_{s,s} = I_n,\quad t > s.$$

Whence it follows that $H_{t,a} = 0$ when $t \geq tr$ and therefore $h_t = 0$.

Consider the generalization to the case when initial conditions are specified only for q of the first components of the state vector Eq. (5.85). Let

$$K_t^{dif} = A_t R_t W_{t+1}^{+} R_t^T C_t^T, \tag{5.87}$$

where

$$R_{t+1} = A_t R_t,\quad R_a = (e_{q+1}, \ldots, e_n), \tag{5.88}$$

$$W_{t+1} = W_t + \Phi_{t,a}^T C_t^T C_t \Phi_{t,a},\ W_a = 0_{(n-q) \times (n-q)}. \tag{5.89}$$

Using Lemma 5.3 with $S_t = 0$ gives

$$H_{t,s} = \Phi_{t,s} - R_t W_t^{-1} \sum_{i=s}^{t-1} R_i^T C_i^T C_i \Phi_{i,s},\quad t \geq t_{tr},\ s < t.$$

It follows from this that $h_t = 0$ when $t \geq tr$.

5.5 FILTERING WITH THE SLIDING WINDOW

It is well known that the KF may have unacceptable estimation accuracy if the system model is not adequate to the real process. In Refs. [60,82] it was proposed the filter with limited memory that is robust in respect to perturbations. The idea is to use a system model that would be adequate to the real process not over the entire interval of observation but only on a limited time moving interval (sliding window). Considered below the DKF with sliding window belongs to this type of algorithm. Besides the fact that it uses a model adequate within the sliding window, the DKF estimate has an important statistical property allowing you to get, unlike the KF, an unbiased estimate for a finite number of steps. This means that for exact measurements it restores the system state in a finite number of steps.

Consider the observation intervals $[t-N,t]$, $t\in T$ (sliding windows), where $N>tr$ and assume that at the moment $t-N$ there is only a priori information about q first components of the vector x_{t-N}. Using the DKF, it is easy to write the following relationships for the evaluating state of the system Eq. (5.1) at the moment t based on a sliding window of the latest N observations

$$\hat{x}^{dif}_{s+1}=A_s\hat{x}^{dif}_s+K^{dif}_s(y_s-C_s\hat{x}^{dif}_s),\ \ \hat{x}^{dif}_{t-N}=(m_a^T,0_{1\times(n-q)})^T, \tag{5.90}$$

where

$$K^{dif}_s=(A_sS_s+A_{1s}R_sW^{+}_{t-N,s+1}R_t^T)C_s^TN_{1s}^{-1}, \tag{5.91}$$

$$\begin{aligned}S_{s+1}&=A_sS_sA_s^T-A_sS_sC_s^TN_{1s}^{-1}C_sS_sA_s^T+V_{2s},\\ S_{t-N}&=block\ diag(\overline{S}_a,0_{(n-q)\times(n-q)}),\end{aligned} \tag{5.92}$$

$$R_{s+1}=A_{1s}R_s,\ \ R_{t-N}=(e_{q+1},\ldots,e_n), \tag{5.93}$$

$$A_{1s}=A_s-A_sS_sC_s^TN_{1s}^{-1}C_s,\ \ N_{1s}=C_sS_sC_s^T+V_{1s}, \tag{5.94}$$

$$W_{s+1,t-N}=W_{s,t-N}+R_s^TC_s^TN_{1s}^{-1}C_sR_s,\ \ W_{t-N,t-N}=0, \tag{5.95}$$

$$s=t-N,t-N+1,\ldots,t-1,t=a+N,a+N+1,\ldots.$$

Let us illustrate the effect of the DKF use with sliding window under parametric uncertainty.

Example 5.1.
Given a sampled model of the aircraft engine [83]

$$x_{t+1}=Ax_t+Bw_t=\begin{bmatrix}0.90305+\delta_t & 0 & 0.1107\\ 0.0077 & 0.9802+\delta_t & -0.0173\\ 0.0142 & 0 & 0.8953+0.1\delta_t\end{bmatrix}x_t+\begin{bmatrix}1\\1\\1\end{bmatrix}w_t$$

taking into account the delay in the measuring channel

$$y_t = Cx_{t-\tau} + D\xi_t = \begin{bmatrix} 1+0.1\delta_t & 0 & 0 \\ 0 & 1+0.1\delta_t & 0 \end{bmatrix} x_{t-\tau} + \begin{bmatrix} 1 \\ 1 \end{bmatrix} \xi_t,$$

where w_t and ξ_t are uncorrelated random processes with zero expectation and variances $E(w_t^2) = 0.02$, $E(\xi^2) = 0.02$, $E(\xi_t^2) = 0.02$, τ, is a delay, and δ_t is an unknown disturbance caused by a temperature change of the turbine.

We want to compare the KF and DKF estimation errors under the action on the engine of the unknown disturbance δ_t. Both filters are constructed without consideration of its action and it is assumed that

$$\delta_t = \begin{cases} 0.1, t \in [50, 100] \\ 0, t \notin [50, 100] \end{cases}.$$

We transform the model to the standard form without delay when $\tau = 1$

$$z_{t+1} = \tilde{A} z_t + \tilde{B} w_t, \quad v_t = \tilde{C} z_t + \tilde{D} \tilde{\xi}_t,$$

where

$$z_t = \begin{bmatrix} x_t \\ Cx_t \end{bmatrix}, \ \tilde{A} = \begin{bmatrix} A & 0 \\ C & 0 \end{bmatrix}, \ \tilde{B} = \begin{bmatrix} B \\ 0 \\ 0 \end{bmatrix},$$

$$\tilde{C} = \begin{cases} 0_{2\times 5}, & t \\ [0_{2\times 3}, I_2], & t \geq 1 \end{cases}, \ \tilde{\xi} = \begin{cases} 0, & t = 0 \\ \xi, & t \geq 1. \end{cases}$$

When constructing the KF we assume that the initial condition for the state vector is known accurately and $x_0 = 0$. The DKF is used in a sliding window mode assuming no a priori information about the initial state vector of the object. Since the dynamics matrix is singular and a priori information is absent then the condition $W_{a,t} > 0$ is only sufficient for the existence of an optimal estimate. This is carried out from the moment $t_{tr} = 3$.

Fig. 5.1 shows the simulation results—the estimation errors $e(t) = \hat{z}_{1t} - z_{1t}$ with the KF and the DKF for $N = 20$ (curves 1 and 2, respectively). It is seen that the errors of the DKF are significantly less than the KF errors on the interval of the perturbation action. In addition, the convergence speed is significantly higher for the DKF compared to the KF after the termination of the disturbance.

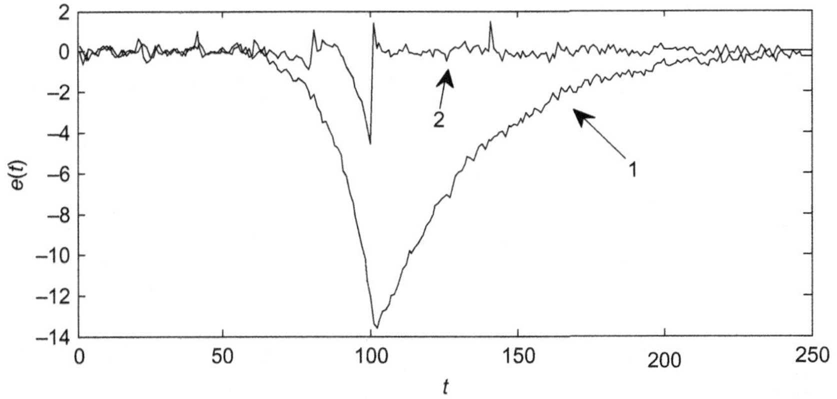

Figure 5.1 Dependencies of estimation errors on the time (curves 1 and 2, respectively).

We note that from the presented results it follows that the DKF with sliding window allows you to get an unbiased estimate when $N > tr$. However, in many cases, also of interest is the statistical spread of the estimates which can be significant even, as was shown in Chapter 3, in the evaluation of linear regression parameters. So the choice of N should be accompanied by correlation analysis of the used algorithm. Such analysis, in problems of the state estimation of the system Eq. (5.1), can be based on the use of the expression (5.82).

5.6 DIFFUSE ANALOG OF THE EXTENDED KALMAN FILTER

Consider a nonlinear dynamical system of the form

$$x_{t+1} = F_t(u_t, x_t) + w_t, \tag{5.96}$$

$$y_t = Z_t(x_t) + \xi_t, \quad t \in T = \{a, a+1, \ldots\}, \tag{5.97}$$

where $x_t \in R^n$ is the state vector, $y_t \in R^m$ is the measured output vector, $u_t \in R^n$ is the known input vector, $F_t(z_t, x_t)$, $Z_t(x_t)$, are the given functions, $w_t \in R^n$ and $\xi_t \in R^m$ are uncorrelated random processes with zero expectation and the covariances $E(w_t w_t^T) = Q_t$, $E(\xi_t \xi_t^T) = R_t$, and the random vector x_a is not correlated with w_t and ξ_t for $t \in T$.

Suppose that conditions A1–A5 or B1–B4 are satisfied and it is required to write the relations for the diffuse extended KF (DEKF) using Theorem 5.1 or 5.3.

Represent the relation for DKF in the following form. Between observations

$$x_{t+1}^{-} = A_t x_t^{+}, \tag{5.98}$$

$$S_{t+1}^{-} = A_t S_t^{+} A_t^T + V_{2t}, \tag{5.99}$$

$$R_{t+1} = A_t(I_n - S_t^{-} C_t^T N_t^{-1} C_t) R_t, \tag{5.100}$$

$$W_{t+1} = W_t + R_t^T C_t^T N_t^{-1} C_t R_t, \tag{5.101}$$

$$N_t = C_t S_t^{-} C_t^T + V_{1t}. \tag{5.102}$$

After receiving the observations

$$K_t^{dif} = S_t^{-} C_t^T N_{1t}^{-1} + (I_n - S_t^{-} C_t^T N_t^{-1} C_t) R_t W_{t+1}^{+} R_t^T C_t^T N_t^{-1}, \tag{5.103}$$

$$x_t^{+} = x_t^{-} + K_t^{dif}(y_t - C_t \hat{x}_t), \tag{5.104}$$

$$S_t^{+} = (I_n - S_t^{-} C_t^T N_t^{-1} C_t) S_t^{-}. \tag{5.105}$$

The initial conditions for the filter are determined by expressions

$$x_a^{+} = (\tilde{m}_a^T, 0_{n-q}^T)^T, \quad S_a^{-} = block\ diag(\overline{S}_a, 0_{(n-q)\times(n-q)}), \tag{5.106}$$

$$R_a = (e_{q+1}, \ldots, e_n), \quad W_a = 0_{(n-q)\times(n-q)}. \tag{5.107}$$

The DEKF is obtained by replacing Eq. (5.98) in

$$x_{t+1}^{-} = F_t(u_t, x_t^{+}) \tag{5.108}$$

and using the linearization

$$A_t = \partial F_t(u_t, x_t^{+})/\partial x_t^{+}, \quad C_t = \partial Z_t(x_t^{+})/x_t^{+}. \tag{5.109}$$

5.7 RECURRENT NEURAL NETWORK TRAINING

Let the recurrent NN be described by Eqs. (4.210) and (4.211). Then the state space model at the training stage will take the form

$$p_{t+1} = \Psi(u_t, p_t, \beta_t) + w_t^p, \tag{5.110}$$

$$\beta_{t+1} = \beta_t + w_t^{\beta}, \tag{5.111}$$

$$\alpha_{t+1} = \alpha_t + w_t^{\alpha}, \tag{5.112}$$

$$y_t = \Phi(p_t)\alpha_t + \xi_t, \quad t = 1, 2, \ldots, N \tag{5.113}$$

with some small random initial conditions p_0, β_0, where $\Psi(z_t, p_t, \beta_t)$, $\Phi(p_t)$, p_t, β_t α_t, are defined by Eqs. (4.212) and (4.213), w_t^p, w_t^{β}, w_t^{α} are uncorrelated artificially added noises with

$$E(w_t^p) = 0, \quad Q^p = E[w_t^p (w_t^p)^T],$$
$$E(w_t^{\beta}) = 0, \quad Q^{\beta} = E[w_t^{\beta} (w_t^{\beta})^T],$$
$$E(w_t^{\alpha}) = 0, \quad Q^{\alpha} = E[w_t^{\alpha} (w_t^{\alpha})^T],$$

At the stage of the prediction the state space model has the form

$$p_{t+1} = \Psi(u_t, p_t, \beta^*) + w_t^p, \tag{5.114}$$

$$y_t = \Phi(p_t)\alpha^* + \xi_t, \quad t = 1, 2, \ldots, N, \tag{5.115}$$

with small random initial conditions p_0, where β^*, α^*, are the parameters estimates obtained by training.

Matrix functions A_t, C_t, in Eq. (5.109) can be easily determined using the relations

$$\frac{\partial}{\partial p_t}\Psi_i(u_t, p_t, \beta_t) = \frac{d\sigma(x)}{dx} a_i, \quad \frac{\partial}{\partial z_t}\Psi_i(u_t, p_t, \beta_t) = \frac{d\sigma(x)}{dx} b_i,$$
$$\frac{\partial}{\partial d}\Psi_i(u_t, p_t, \beta_t) = \frac{d\sigma(x)}{dx}, \quad i = 1, 2, \ldots, q,$$
$$\frac{\partial}{\partial c_j}\Phi_j(p_t) = p_t, \quad j = 1, 2, \ldots, m$$

where $x = a_i^T p_{t-1} + b_i^T u_t + d_i$.

Example 5.2.

Let a plant be described by the equations system [84,85]

$$x_{1t+1} = \frac{x_{1t} + 2x_{2t}}{1 + x_{2t}^2} + u_t, \quad x_{2t+1} = \frac{x_{1t}x_{2t}}{1 + x_{2t}^2} + u_t,$$

$$y_t = x_{1t} + x_{2t}, \quad \in t = 1, 2, \ldots, N. \tag{5.116}$$

To simulate the plant we used the recurrent NN Eqs. (4.210) and (4.211) with

$$q = 10, m = 1, N = 900, \quad M = 100 \text{ (number of test points)},$$

$$Q = block\ diag(Q^p, Q^\beta, Q^\alpha) = 0.001I_{140}, \quad R = 1, \quad P_0, \quad \beta_0, u_t \sim 0.001 \times rand(-1, 1).$$

The ability to generalization of the recurrent NN additionally checked using the harmonic input signal

$$u_t = \sin(2\pi/t), \quad t = 1, 2, \ldots, N.$$

The simulation results are shown in Figs. 5.2–5.4.

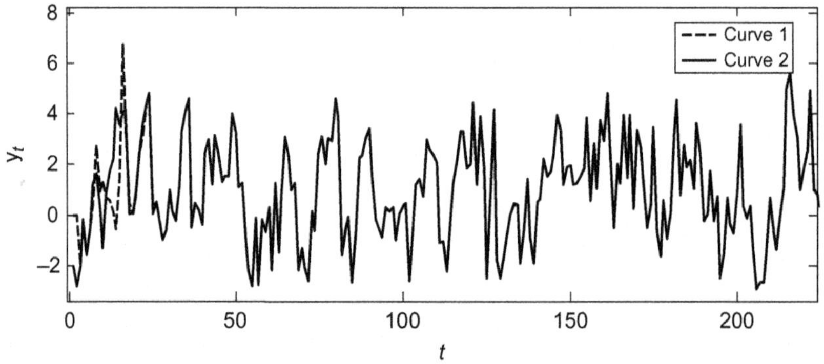

Figure 5.2 Dependencies of the DEKF output (curve 1) and the output plant (curve 2) on the number observations in the training set.

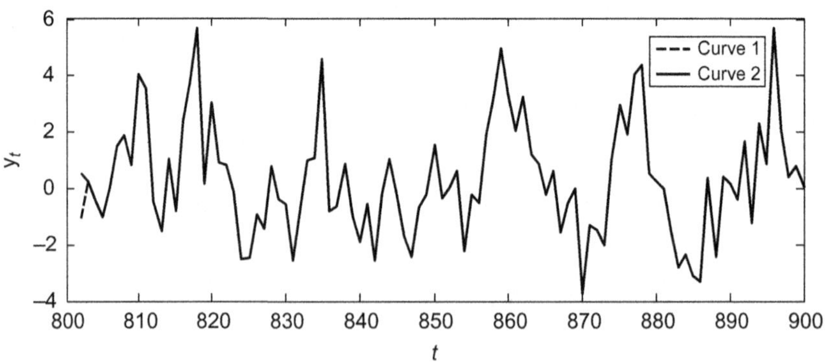

Figure 5.3 Dependencies of the DEKF output (curve 1) and the output plant (curve 2) on the number observations in the testing set.

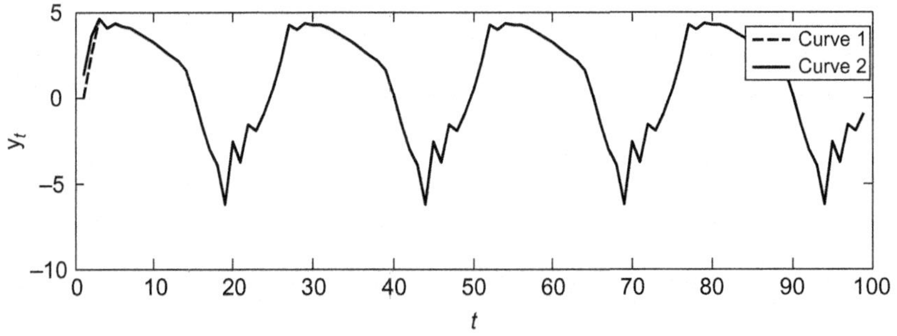

Figure 5.4 Dependencies of the DEKF output (curve 1) and the output plant (curve 2) on the number observations in the testing set with the input signal $u_t = \sin(2\pi/t)$.

5.8 SYSTEMS WITH PARTLY UNKNOWN DYNAMICS

Consider the problem of estimation of the system state Eq. (5.96) when its right-hand part contains some level of uncertainty

$$x_{t+1} = F_t(u_t, x_t) + \varepsilon(z_t, x_t) + w_t, \tag{5.117}$$

$$y_t = Z_t(x_t) + \xi_t, \quad t = 1, 2, \ldots, N, \tag{5.118}$$

where $\varepsilon(z_t, x_t)$ is the difference between the total and the approximate model of the plant.

Assume that $\varepsilon(z_t, v_t)$ is a continuous function. We can use a perceptron with one hidden layer and linear activation function in the output or the RBNN Sugeno fuzzy system described by separable regression

$$\varepsilon(z_t, v_t) = \Phi(z_t, \beta)\alpha, \tag{5.119}$$

for the approximation of $\varepsilon(z_t, v_t)$ on compact sets, where $z_t = (x_t^T, u_t^T)^T$. Thus, the problem reduces to the simultaneous assessment of the state Eq. (5.117) and the parameters (if they are not known) with the help of measured outputs using the DEKF.

Example 5.3.
Let a plant be described by the equations

$$x_{1t+1} = \frac{x_{1t} + 2x_{2t}}{1 + x_{2t}^2} + u_t + w_t = f_{1t}, \tag{5.120}$$

$$x_{2t+1} = \frac{x_{1t}x_{2t}}{1 + x_{2t}^2} + u_t + w_t = f_{2t}, \tag{5.121}$$

$$y_t = x_{1t} + x_{2t} + \xi_t, \quad t = 1, 2, \ldots, N, \tag{5.122}$$

where $u_t = 10\sin(2\pi/t)$, $w_t \sim N(0, 0.7)$, $\xi_t \sim N(0, 0.1)$.

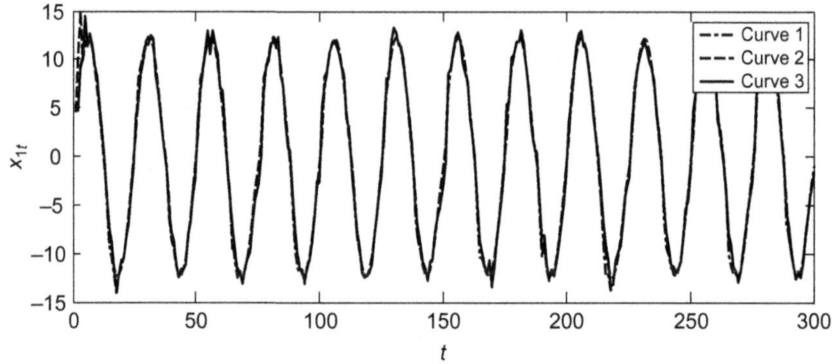

Figure 5.5 Dependencies of x_{1t} (curve 1), estimates of x_{1t} for a known (curve 2) and unknown f_{2t} (curve 3) on the number observations.

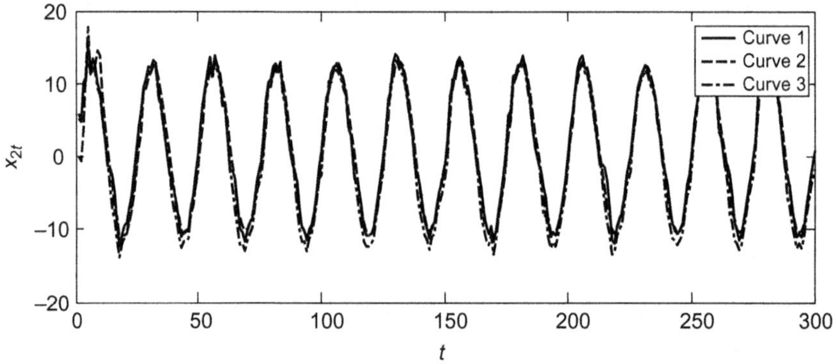

Figure 5.6 Dependencies of x_{2t} (curve 1), estimates of x_{2t} for a known (curve 2) and unknown f_{2t} (curve 3) on the number observations.

Suppose that f_{2t} is an unknown function and we use for the approximation perceptron with one hidden layer

$$f_{2t} = \Phi(z_t, \beta)\alpha = (\sigma(a_1 z_t + b_1), \sigma(a_2 z_t + b_2), \ldots, \sigma(a_r z_t + b_r))\alpha,$$

where $z_t = (x_{1t}, x_{2t}, u_t)^T$, a_k, b_k, $k = 1, 2, \ldots, r$ are weights and biases, respectively, $\sigma(\cdot)$ is the sigmoid function. Let the weights and biases be selected from a uniform distribution:

$$a_k, \ \ b_k \sim 0.01 \times rand(-1, 1), \quad k = 1, 2, \ldots, r$$

and the parameter α be not known. The simulation results for three neurons are shown in Figs. 5.5 and 5.6.

CHAPTER 6

Applications of Diffuse Algorithms

Contents

6.1 IDENTIFICATION OF THE MOBILE ROBOT DYNAMICS

The mobile robot (MR) Robotino shown in Fig. 6.1 is an autonomous mobile platform with three "omnidirectional" wheels [86]. Movement of the robot is carried out with the help of three DC motors whose axes are arranged at angles of 120° to each other.

The speed of the shaft rotation of each motor is transmitted to the axis of the corresponding wheel via the gearbox with a gear ratio of 16:1. The robot has a set of commands that allow setting and measuring the angular speed of the motor shaft rotation

$$\omega^r(t) = (\omega_1^r(t), \omega_2^r(t), \omega_3^r(t))^T, \; \omega^{out}(t) = (\omega_1^{out}(t), \omega_2^{out}(t), \omega_3^{out}(t))^T.$$

Measuring of angular velocities is carried out using an incremental tachometer. Robotino runs under an embedded operating system Linux and work offline is provided by batteries.

The choice of the control that moves MR along a predetermined path is one of the major tasks arising in its design. Assume that the rotational speeds of motor shafts perfectly follow the input signals, the used sensors have high accuracy and the wheels slip is absent. Then the solution of this task can be obtained only on the basis of the kinematics equations. However, the neglect of the influence of the dynamics may limit the ability of the MR in real situations. We will consider the problem of mathematical models constructing of the MR from the experimental data linking the values of given angular velocities $\omega^r(t)$ with their measured

Diffuse Algorithms for Neural and Neuro-Fuzzy Networks.
DOI: http://dx.doi.org/10.1016/B978-0-12-812609-7.00006-8

Figure 6.1 Robotino exterior view.

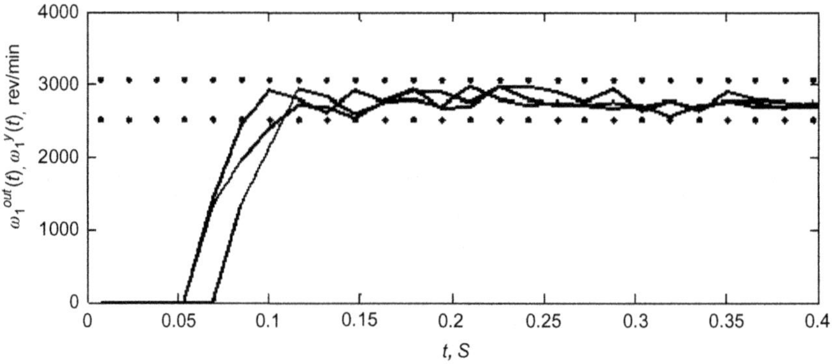

Figure 6.2 Transient performance of engines.

values $\omega^{out}(t)$. An external control program in C that, with the help of appropriate commands, sets the desired angular velocities of the robot $\omega^{r}(t)$ and receives the relevant measurements of speeds $\omega^{out}(t)$ was written.

According to the results of experimental studies it was found that the MR frequency bandwidth does not exceed 8 Hz. Transients for each engine shown in Fig. 6.2 were obtained and the sampling was chosen with the help of these. Besides it was also shown that the engine interference is weak and therefore models of the MR can be built independently for each input and output.

Test signals representing the Gaussian white noise were generated for each input. Noise intensities were chosen according to the rule "3 sigma" to ensure the motor angular shafts rotation in the range of $\omega_i^r(t) \in [-\omega_{\max}^r, \omega_{\max}^r], i = 1, 2, 3$ (rev/min). Furthermore, the generated input signals have been previously passed through the Batervord low-pass filter with bandwidth of the MR equal to 8 Hz.

Figs. 6.3–6.5 show fragments of the input and the corresponding output of the angular speeds of the motors. Clearly a difference is visible between inputs and outputs due to a delay, dynamics of the plant, and the possible influence of nonlinearities.

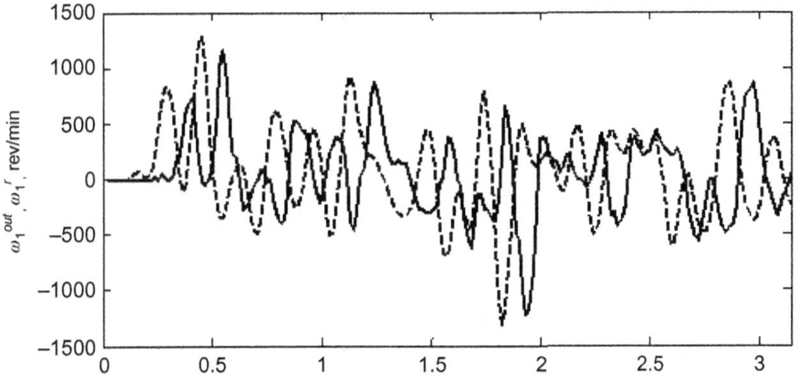

Figure 6.3 Fragment input–output data for identification; $\omega_1^{out}(t)$ is shown by a solid line, $\omega_1^r(t)$ by a dashed line.

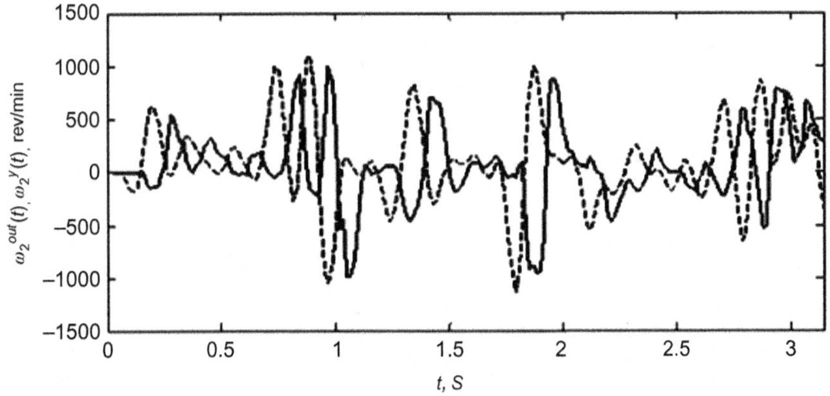

Figure 6.4 Fragment input–output data for identification; $\omega_2^{out}(t)$ is shown by a solid line, $\omega_2^r(t)$ by a dashed line.

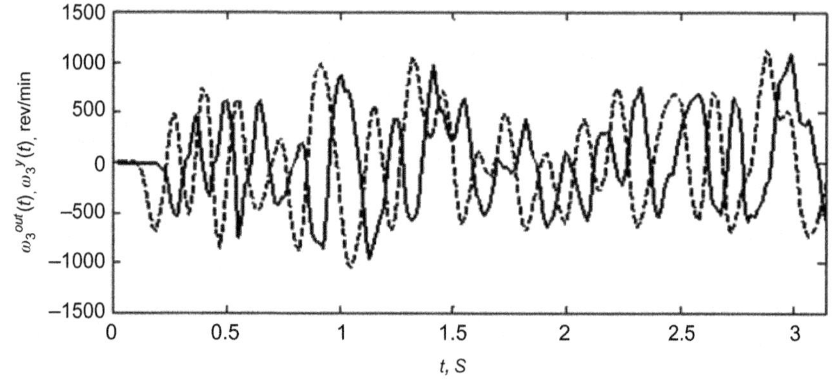

Figure 6.5 Fragment input–output data for identification; $\omega_3^{out}(t)$ is shown by a solid line, $\omega_3^{r}(t)$ by a dashed line.

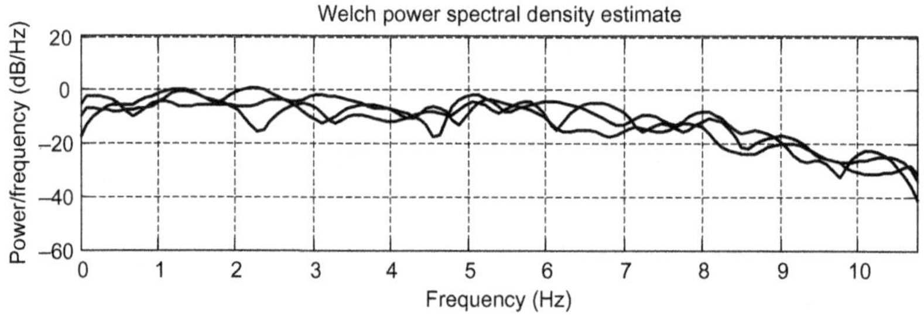

Figure 6.6 Spectral densities of the input signals.

Identification of models was carried out using the sample of 1000 points for each input and output. Obtained data were divided into two parts—training (850 points) and testing (150 points). The first part was used to build models and the second to test them. To obtain measurements equally spaced in time the input/output data were linearly interpolated with the step 0.0156 s that coincides with the median of the tact distribution. Fig. 6.6 shows the spectral densities of the input signals.

We begin with a description of the identification results using a linear model of autoregression and moving average

$$y_t + a_1 y_{t-T} + \cdots + a_q y_{t-qT} = b_1 u_{t-kT} + \cdots + b_l u_{t-(k+l)T} + w_t,$$

where $u_t \in R$, $y_t \in R$ are an input and an output, respectively, w_t is a centered white noise, a_i, $i = 1, 2, \ldots, q$, b_j, $j = 1, 2, \ldots, l$, k are model parameters to be estimated from experimental data, and T is a discreteness tact.

This model describes the dependence between the given and measured values of angular velocities of the motor.

Estimates of the unknown parameters $a_i,\ i=1,2,\ldots,q,\ b_j,\ j=1,2,\ldots,l$ were obtained by minimizing the sum of squared residuals between measured values of the plant output $y(t_i)$ and the model $\hat{y}(t_i),\ i=1,2,\ldots,N$ for the different values of $q,\ l,\ k$.

The adequacy of the constructed models was tested by estimating the determination coefficients

$$R^2=\left(1-\frac{\sum_{i=1}^{N}(y(i)-\hat{y}(i))^2}{\sum_{i=1}^{N}(y(i)-\tilde{y})^2}\right)\times 100\%$$

for each output and using the statistical properties of residues and their correlations with inputs, where $\hat{y}(i)$ is the value predicted by the model and $\tilde{y}$ is the average of the values obtained by the sample.

The accuracy of the data approximation was estimated by the root mean square error (RMSE)

$$RMSE=\left(1/N\sum_{i=1}^{N}(y(i)-\hat{y}(i))^2\right)^{1/2}.$$

The models' order and the delay were chosen so as to ensure the adequacy of the models by comparing the determination coefficients of different models with ranges of possible values $q=1,2,3;\ l=1,2,3;\ k=1,2,\ldots,7$.

Figs. 6.7–6.9 show the simulation results of the obtained models with parameters $[q,l,k]$: $[3,2,7],[3,2,5],[1,3,6]$ for each of the motors. The RMSE of the predictions by one step for each of the motors were

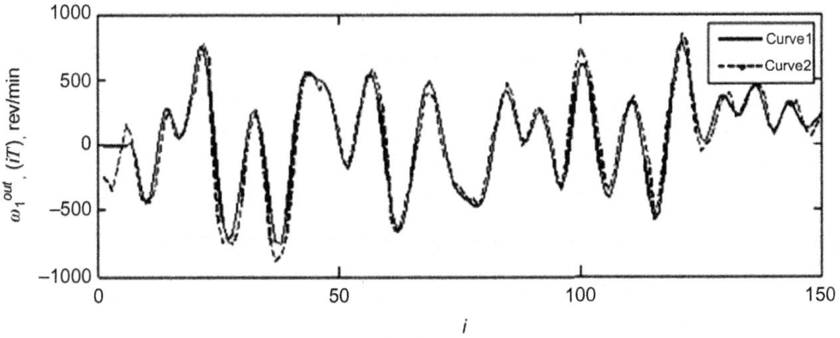

Figure 6.7 Dependences of the angular speed of the first motor (curve 1) and its prediction (curve 2) on the number of observations.

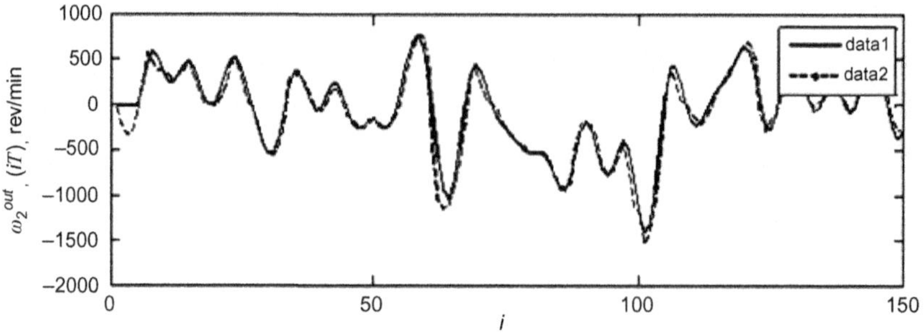

Figure 6.8 Dependences of the angular speed of the second motor (curve 1) and its prediction (curve 2) on the number of observations.

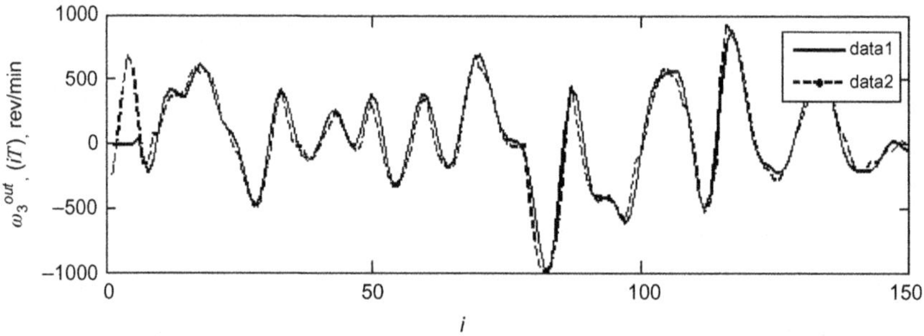

Figure 6.9 Dependences of the angular speed of the third motor (curve 1) and its prediction (curve 2) on the number of observations.

109.4, 112.1, and 117.8 rev/min and the determination coefficients were 79.9%, 81.6%, and 78.5%, respectively.

Fig. 6.10 shows the correlation functions of residues and cross-correlation function inputs with residues.

Let's try to reduce received values of the RMSE. Suppose that the plant is described by a nonlinear autoregressive and moving average (1.14) which is constructed on the basis of the perceptron with one hidden layer, the sigmoid FA in hidden layer and linear in the output, where $z_t = (y_{t-1}, \ldots, y_{t-qT}, u_{t-kT}, \ldots, u_{t-(k+l)T})^T$.

We put for all three engines $q = 3$, $l = 2$, $k = 7$ and the number of neurons in the hidden layer equal to 10 and we estimate the weights and biases of the Neural Network (NN) from observations of input/output pairs, using the iterative diffuse training algorithm (DTA). The forgetting factor is determined from the relationship $\lambda_i = \max\{1 - 0.05/i, 0.99\}$, $i = 1, 2, \ldots, M$, $M = 10$ is the iterations number.

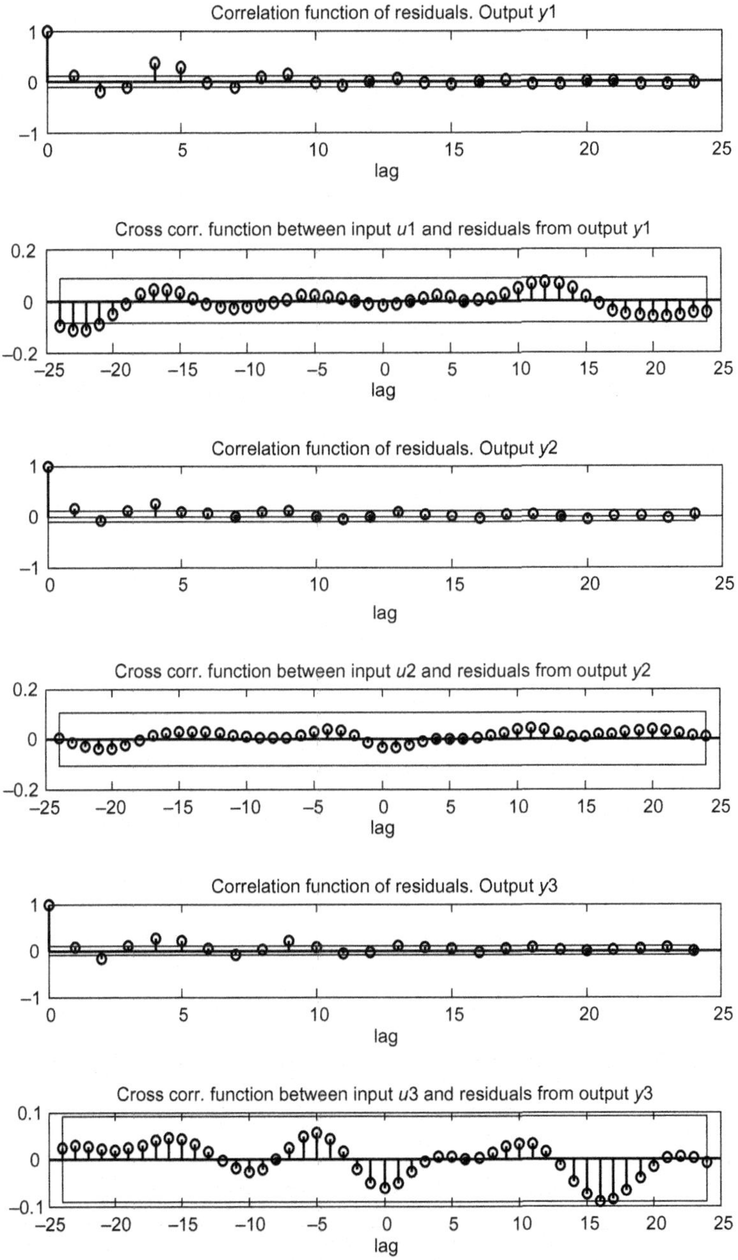

Figure 6.10 Correlation functions of the residues and cross-correlation function inputs with the residues.

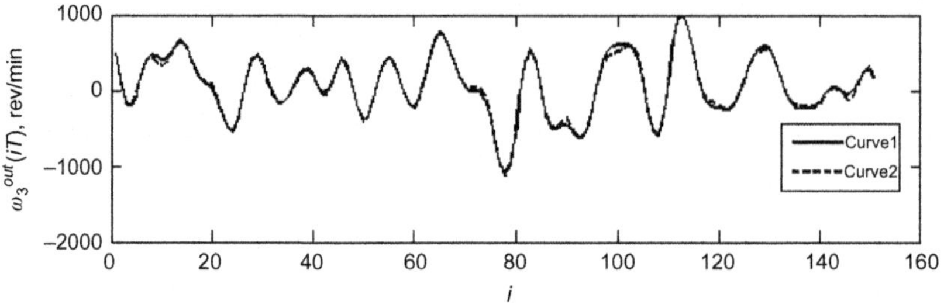

Figure 6.11 Dependences of the angular speed of the first motor (curve 1) and its prediction (curve 2) on the number of observations.

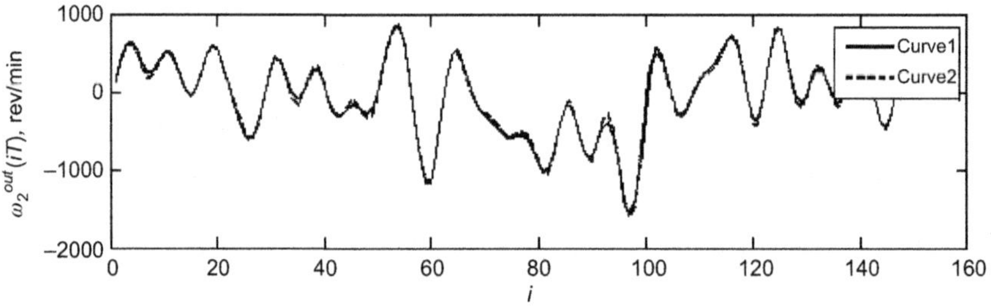

Figure 6.12 Dependences of the angular speed of the second motor (curve 1) and its prediction (curve 2) on the number of observations.

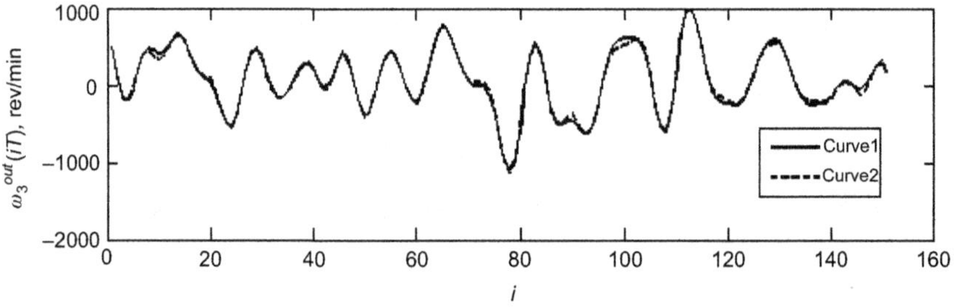

Figure 6.13 Dependences of the angular speed of the third motor (curve 1) and its prediction (curve 2) on the number of observations.

Figs. 6.11–6.13 show the simulation results for the test sample of the obtained models for each of the engines. Here graphs with numbers 1 are the experimental values of the angular velocities, with the numbers 2 being predictions by one step found with the help of the models, using the testing set. The RMSE of the predicted values by 1 step are significantly less than for the linear autoregressive and moving average models—40.9, 45.5, 43.4 rev/min for each of the motors, respectively.

Increasing of the approximation accuracy is noticeable even in the visual comparison with the graphs shown in Figs. 6.7–6.9. The weights and the biases of the hidden layer are assumed initially to be uniformly distributed random variables on the intervals [1.1] and [0.1], respectively.

6.2 MODELING OF HYSTERETIC DEFORMATION BY NEURAL NETWORKS

Hysteresis relationships between different variables appear in many engineering applications.

However, detailed modeling of systems with hysteresis, using physical laws, is usually a difficult task and the resulting models are often too complex to be used in practical applications. Therefore, various alternative approaches that are not the result of deep analysis of the physical behavior of the system, but combining some physical understanding of hysteresis with models such as input–output, were proposed [87]. As an example, numerous well-known and successful applications of methods for constructing of hysteresis models using NNs and fuzzy logic in the problems of ferromagnetism and mechanics are proposed in [88,89]. The idea of their application is based on one of the main manifestations of hysteresis—dependence of the system output $y(t)$ at the moment t on the previous and current values of the input $u(t_p), t_p \leq t$. It is shown that the dependence of this type can be simulated by a dynamic system with a specially selected state, input and output and the NN is used to approximate its right side. Note that it seems that the first results where a NN was used to model the deformation dependence on the effort in the concrete plate are presented in [90,91]. In this section we show that the iterative DTA can be used for the NN training to simulate a hysteresis.

We use the experimental data shown in Fig. 6.14. Here $F(\delta)$ is a mechanical force (load) applied to a detail and δ is deformation caused by $F(\delta)$.

In the experiments randomly implemented dependences for four levels of the force are equal to 0, 90, 140, and 250 kg. Figs. 6.15 and 6.16 show the dependence of the force and the corresponding deformation on the observations number. The frequency of data pickup is 20 Hz. It is seen that the error in putting, for example, 250 kg level can reach 23 kg. This explains the variability of the hysteresis curves shown in Fig. 6.14.

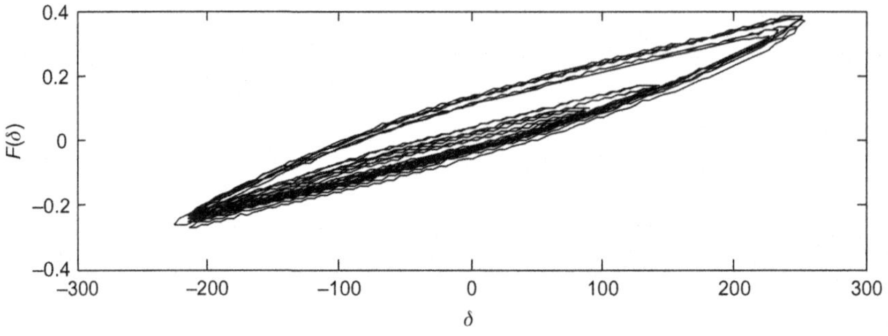

Figure 6.14 The dependence of the deformation (mm) from the applied force (kg).

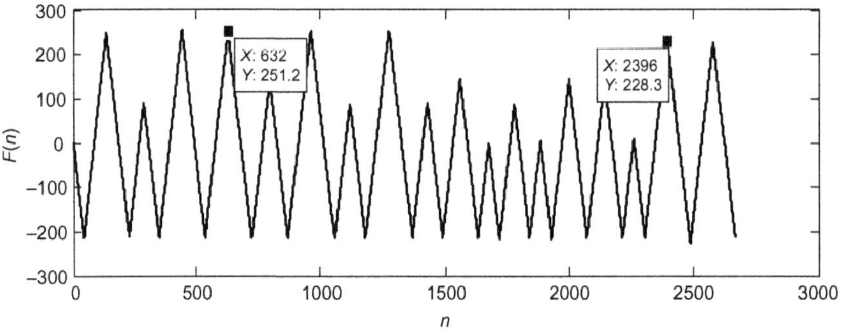

Figure 6.15 The dependence of force $F(n)$ (kg) on the number of observations n.

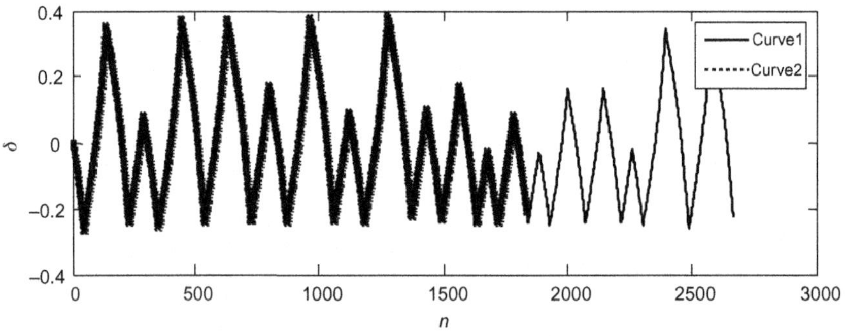

Figure 6.16 Dependences of deformation δ (mm) (curve 1) and its approximation (curve 2) on the number of observations n in the training set.

Observing the behavior of the curve in Fig. 6.14 it is easy to see that the knowledge of force value is not enough to uniquely identify the deformation. This can be corrected by the introduction of additional variables that take into account the background of the process (the change of the deformation under the influence of the varying force). This may be, for example, the force and the deformation values at some point prior to the current time. In more general cases the deformation value at the moment t can be defined by the difference equation

$$\delta(t) = f(\delta(t-\Delta), \ldots, \delta(t-a\Delta), F(t-d), \ldots, F(t-d-b\Delta)), \quad (6.1)$$

where f is some function, Δ is a time sample rate, and a, b, d are some positive integers.

The difference Eq. (6.1) is a nonlinear model of autoregressive and moving average by means of which we try to describe the hysteresis phenomen. Therefore, the problem of constructing a hysteresis model is reduced to the identification of the discrete dynamic system. We rely on the following common approach to its decision:

1. The function f is sought in the class of the following parameterized dynamic dependencies

$$\begin{aligned}\delta(t) &= \Phi(W, \delta(t-\Delta), ..., \delta(t-a\Delta), F(t), ..., F(t-b\Delta)) \\ &= \Phi(W, Z(t)),\end{aligned} \quad (6.2)$$

where W is a vector of unknown parameters and $z(t) = (\delta(t-\Delta), \ldots, \delta(t-a\Delta), F(t), \ldots, F(t-b\Delta))^T$ is a input vector. The function $\Phi(W, z(t))$ should be a universal approximator, in the sense that any continuous function can be reproduced by $\Phi(W, z(t))$ with arbitrarily high accuracy.

2. A set of input/output data $\{F(t), \delta(t)\}, t = 0, \Delta, \ldots, N\Delta$ is divided into two groups, training $t = 0, \Delta, \ldots, N_1\Delta$ and testing $t = (N_1+1)\Delta, (N_1+2)\Delta, \ldots, N\Delta$ (the latter is only used to verify the adequacy of the model).

3. It is fixed the permissible range of a, b, d parameters and W is selected from this range minimizing the quadratic criterion

$$Q(W) = 1/N_1 \sum_{i=q}^{N_1} (\delta(i\Delta) - \Phi(W, z(i\Delta))^2. \quad (6.3)$$

Then the values of criteria for various a, b are compared, and the vector W minimizing the criteria is chosen.

4. The model is tested in respect to the reproduction accuracy of the testing data, residues correlation values, and residues correlation values with inputs.

Figs. 6.17–6.22 show the modeling results of the hysteretic relationship shown in Fig. 6.14 with the help of the nonlinear difference equation

$$\delta(t) = \Phi(W, \delta(t-\Delta), F(t), F(t-\Delta)), t \in T = \{0, \Delta, \ldots, N\Delta\}. \quad (6.4)$$

Eq. (6.4) determines uniquely the deformation $\delta(t)$ at the points $t \in \{\Delta, 2\Delta, \ldots\}$ for the values $F(t), t \in \{\Delta, 2\Delta, \ldots\}$ if you set the values $F(0)$, $F(-\Delta)$ and the deformation $\delta(-\Delta)$.

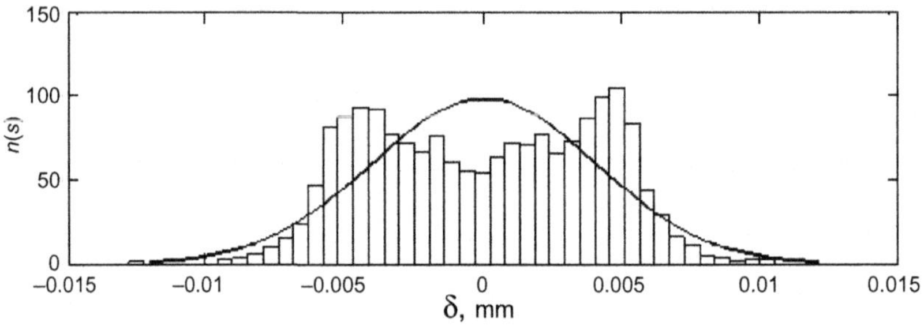

Figure 6.17 Histogram of residues on the training set.

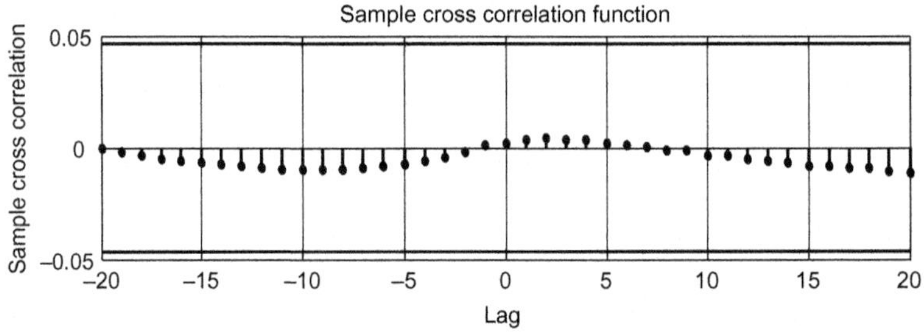

Figure 6.18 Correlation residues with $F(t)$ on the training set.

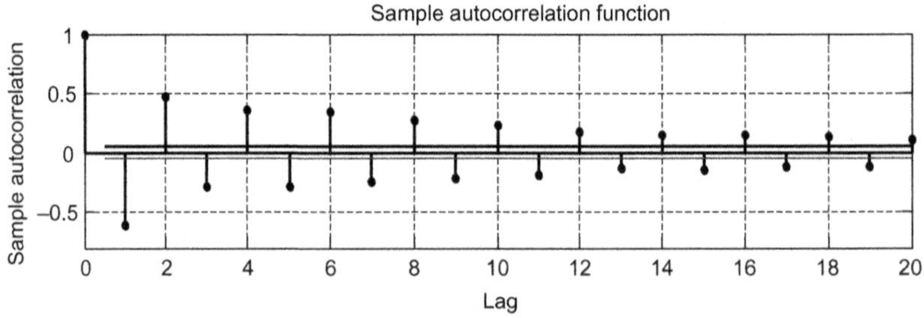

Figure 6.19 Correlation residues on the training set.

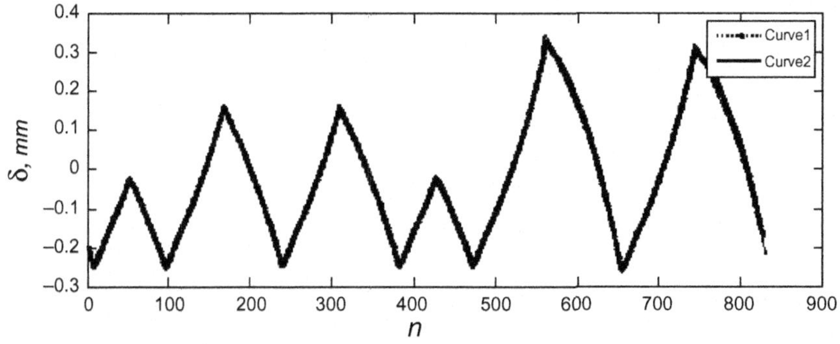

Figure 6.20 Dependences of deformation δ (mm) (curve 1) and its approximation (curve 2) on the number of observations n on the testing set.

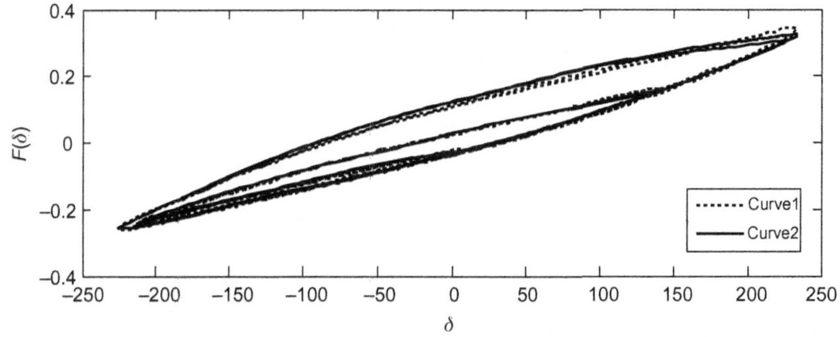

Figure 6.21 Dependences of the deformation (curve 2) and approximating curve (curve 1) (mm) from the applied force (kg) on the test set.

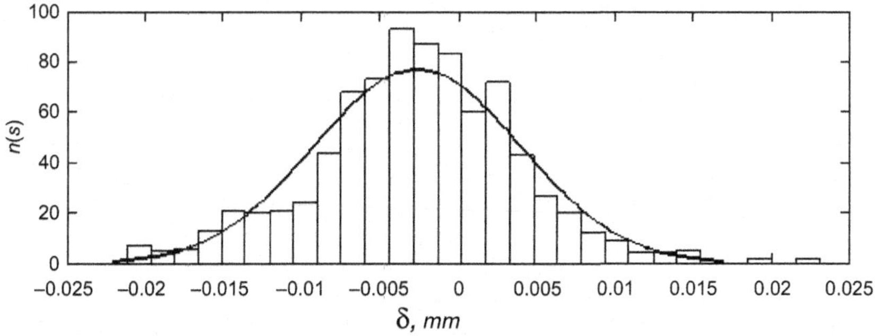

Figure 6.22 Histogram of residues on the testing set.

To approximate the right-hand side in Eq. (6.4) we use a perceptron with one hidden layer and the linear output activation function (AF), and five neurons in the hidden layer. The size of the training set is 1836, the test is 824. The forgetting factor is determined by expression $\lambda_i = \max\{1 - 0.05/i, 0.99\}$, $i = 1, 2, \ldots, 100$.

Of interest is also the question of how the use of such complex mathematical structures as NNs to construct the hysteresis curves shown in Fig. 6.14 can be justified. It should be noted that the simulation results show that a linear model of the form

$$\delta_m(t) = a_1\delta_m(t-\Delta) + b_1F(t) + b_2F(t-\Delta)$$

can't approximate even the points of the training set. A similar result was obtained by using the state space model

$$x(t+\Delta) = Ax(t) + BF(t) + \xi(t), \delta(t) = Cx(t) + \zeta(t)$$

and the subspace identification method for evaluation of A, B, C.

6.3 HARMONICS TRACKING OF ELECTRIC POWER NETWORKS

Evaluation of the periodic signal harmonics amplitudes is an extremely important problem in the networks of electricity consumption. This is due to the fact that the analysis of the most electrical equipment is based on the assumption that the load current can be represented by a superposition of harmonics with frequencies that are multiples of the fundamental frequency. Besides, real-time estimation algorithms of harmonics amplitudes are needed for the harmonic correction. To solve this problem recursive algorithms [92–94] are used. Their work is based on the following idea. Suppose that we have a priori information about the number of required harmonics for reproduction of the observed load currents with a given accuracy, then estimates of the harmonics amplitudes (Fourier coefficients) are proposed to find through a recursive minimization of the squared prediction errors at one step between the load current and model values (the Widrow-Hoff algorithm) or using the RLSM. At the same time, the rate of convergence of the Widrow-Hoff algorithm may be unsatisfactory and the RLSM can diverge under the action of perturbations and changing loads.

Compare the features of the Widrow–Hoff algorithm and the DTA with sliding window on numerical examples with real data.

Let the mathematical model of the signal have the form

$$y_t = A_0 + \sum_{i=1}^{M}[A_i \sin(2\pi fit) + B_i \cos(2\pi fit)] + \xi_t = C_t\alpha + \xi_t, \qquad (6.5)$$

where $f = 60$ Hz is the fundamental frequency, $M = 23$, $\alpha = (A_0, A_1, \ldots, A_M, B_1, \ldots, B_M)^T$,

$$C_t = (1, \sin(2\pi ft), \ldots, \sin(2\pi fMt), \cos(2\pi ft), \ldots, \cos(2\pi fMt)).$$

It is needed to evaluate recursively values of the amplitudes of harmonics in Eq. (6.5), using the Widrow–Hoff algorithm

$$\alpha_t = \alpha_{t-1} + \eta C_t(y_t - C_t\alpha_{t-1}), \quad \alpha_0 = 0, \quad \eta = 0.005$$

and the DTA with sliding window. Tact data pickup T over the period of the fundamental frequency is 256 and the prediction horizon of the DTA h with sliding window is 300.

Figs. 6.23–6.25 show a fragment of the realization with an impulse disturbance, evaluation of harmonics amplitudes obtained by the Widrow–Hoff algorithm (curves 1) and the DTA with sliding window (curves 2), where $y_t = y_{iT}$, $i = 1,2,\ldots$, A is shortened to the ampere. It is evident that the DTA substantially exceeds the Widrow–Hoff algorithm in speed.

Figs. 6.26–6.28 show a fragment of the realization with the varying load, evaluation of the amplitudes of the harmonics produced by the Widrow–Hoff algorithm (curves 1), and the DTA with sliding window (curve 2) in this case. It is seen that the DTA in this case substantially exceeds the Widrow–Hoff algorithm in speed.

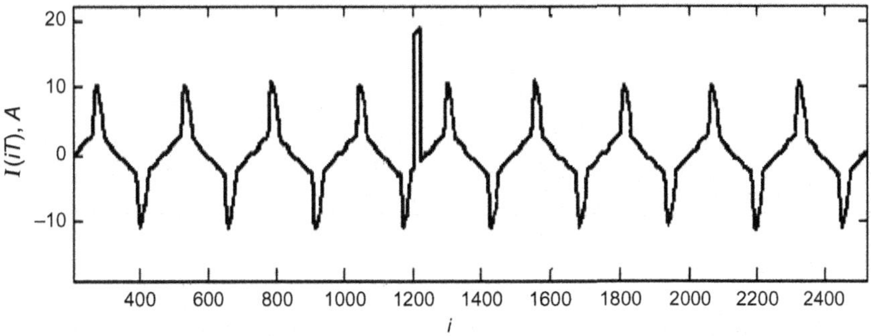

Figure 6.23 The dependence of the current value on the number of observations.

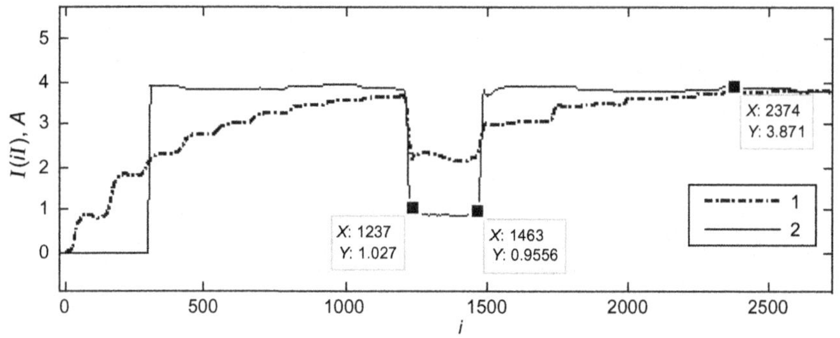

Figure 6.24 Dependencies of harmonic amplitude estimates A_1 on the number of observations under the impulse noise action. The Widrow–Hoff algorithm (curve 1) and the DTA with sliding window (curve 2).

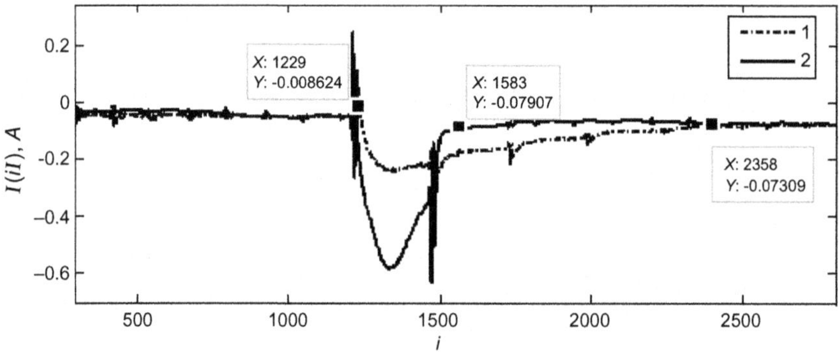

Figure 6.25 Dependencies of harmonic amplitude estimates A_{23} on the number of observations under the impulse noise action. The Widrow–Hoff algorithm (curve 1) and the DTA with sliding window (curve 2).

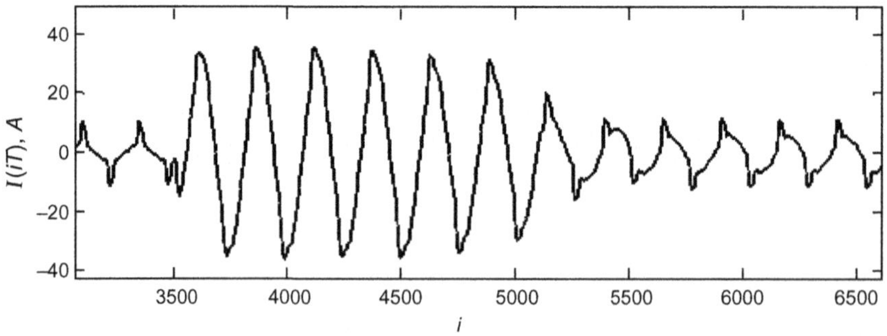

Figure 6.26 The dependence of the signal value on the number of observations.

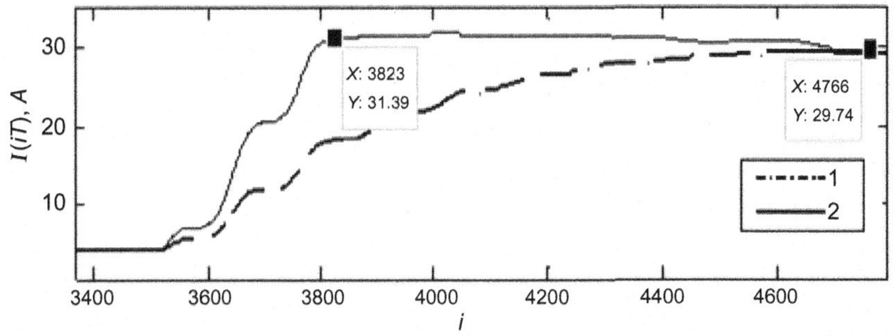

Figure 6.27 Dependencies of harmonic amplitude estimates A_1 on the number of observations while the varying load. The Widrow–Hoff algorithm (curve 1) and the DTA with sliding window (curve 2).

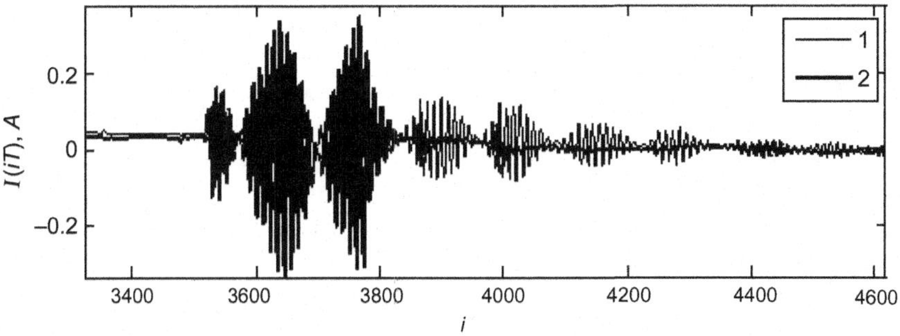

Figure 6.28 Dependencies of harmonic amplitude estimates A_{23} on the number of observations while the varying load. The Widrow–Hoff algorithm (curve 1) and the DTA with sliding window (curve 2).

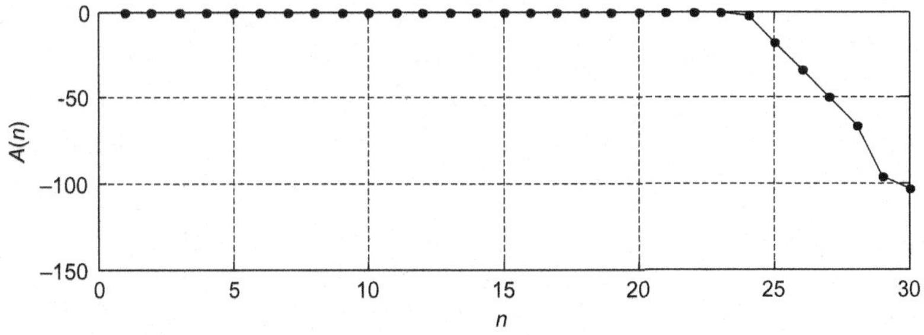

Figure 6.29 The dependence of the amplitude–frequency characteristics of the filter on the harmonic number.

Signals are preprocessed by a low-pass elliptic filter of the tenth order with the delay payment. The amplitude–frequency characteristic of the filter is shown in Fig. 6.29.

The problem of simultaneous estimation of the harmonics amplitudes and the fundamental frequency change in electric current monitoring in networks is also of considerable interest.

Figs. 6.30–6.32 show the results of the simultaneous estimation of harmonics amplitudes, and the fundamental frequency period under the action impulse noise (Fig. 6.23) using nonlinear DTA.

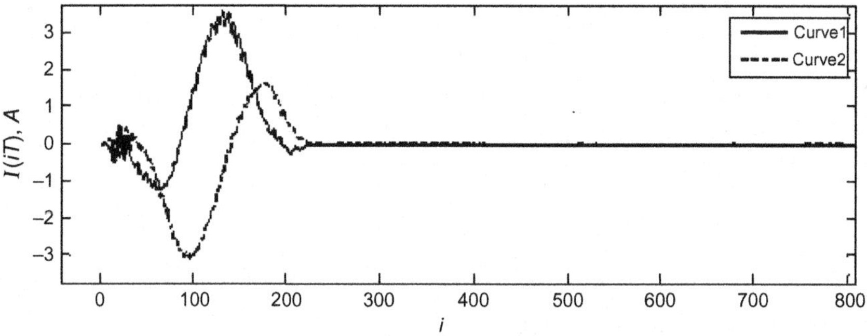

Figure 6.30 Dependencies of harmonics amplitudes estimates A_1 (curve 1) and B_1 (curve 2) on the observations number under the impulse noise action.

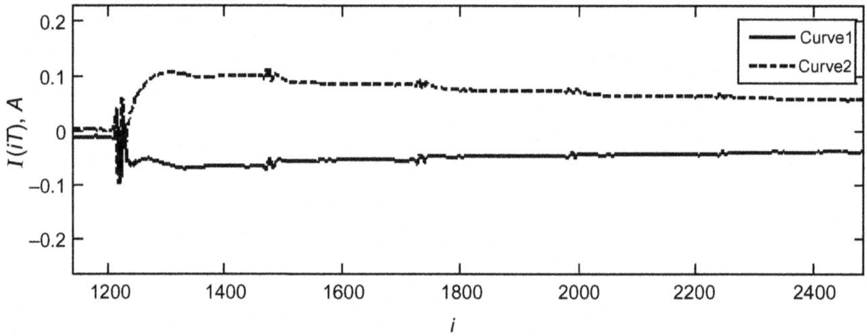

Figure 6.31 Dependencies of harmonics amplitudes estimates A_{23} (curve 1) and B_{23} (curve 2) on the observations number under the impulse noise action.

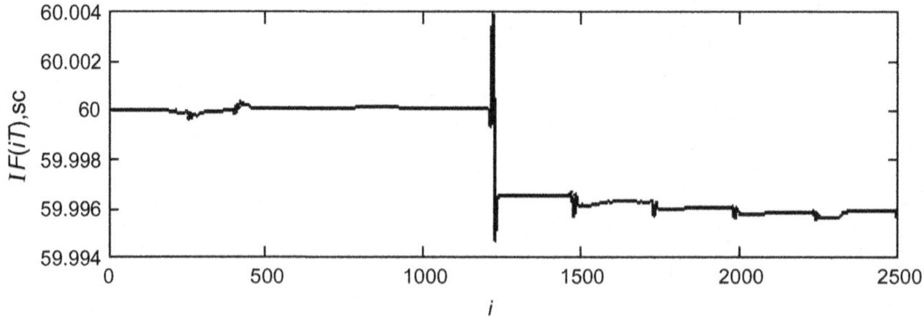

Figure 6.32 The dependence of the estimation of the fundamental period on the observations number under the impulse noise action.

GLOSSARY

NOTATIONS

A^T transposition of matrix A
A^{-1} inverse of matrix A
A^+ pseudoinverse of matrix A
$A > 0$ positive definite matrix A
$A \geq 0$ positive semidefinite matrix A
$A = diag(a_1, ..., a_n)$ diagonal matrix
$A = block\ diag(A_1, ..., A_n)$ block-diagonal matrix
$trace(A)$ trace of matrix A
$detA$ determinant of matrix A
$0_{n \times m}$ zero $n \times m$ matrix
I_n identity $n \times n$ matrix
e_i i-th unit vector of dimension n
R^n n-dimensional linear space over the field of real numbers
$R^{n \times m}$ set of $n \times m$ matrices
$\otimes$ direct product of two matrices
$rank(A)$ rank of matrix A
$E(\xi)$ expectation of vector ξ
$||.||$ Euclidean vector norm

ABBREVIATIONS

AF activation function
DEFK diffuse extended Kalman filter
DFK diffuse Kalman filter
DTA diffuse training algorithm
EKF extended Kalman filter
ELM extreme learning machine
GNM Gauss–Newton method
KF Kalman filter
LSM least-square method
MF membership function
NFS neuro-fuzzy system
NN neural network

RBNN radial basic NN
RLSM recursive least-square method
RNN recurrent NN
SR separable regression

REFERENCES

[1] Golub GH, Pereyra V. The differentiation of pseudoinverses and nonlinear least squares problems whose variables separate. SIAM J Number Anal 1973;10:413–32.
[2] Golub GH, Pereyra V. Separable nonlinear least squares: the variable projection method and its applications. Inverse Probl 2003;19(2):R1–26.
[3] Pereyra V, Scherer G, Wong F. Variable projections neural network training. Math Comput Simul 2006;73(1–4):231–43.
[4] Sjoberg J, Viberg M. Separable non-linear least-squares minimization and possible improvements for neural net fitting. In: IEEE workshop in neural networks for signal processing, FL, USA; 1997, p. 345–354.
[5] Jang R. Fuzzy modeling using generalized neural networks and Kalman filter algorithm. In Proceeding of the Ninth National Conference on Artificial Intelligence (AAAI-91), p. 762–7, July, 1991.
[6] Jang R, Sun C, Mizutani E. Neuro-Fuzzy and Soft Computing: A Computational Approach to Learning and Machine Intelligence. Englewood Cliffs, NJ: Prentice Hall; 1997.
[7] Huang GB, Zhu QY, Siew CK. Extreme learning machine: theory and applications. Neurocomputing 2006;70:489–501.
[8] Rong HJ, Huang GB, Sundararajan N, Saratchandran P. Online sequential fuzzy extreme learning machine for function approximation and classification problems. IEEE Trans Syst Man Cybern B Cybern 2009;39(4):1067–72.
[9] Kaufman L. A variable projection method for solving separable nonlinear least squares Problems. BIT 1975;15:49–57.
[10] Haykin S. Neural Networks and Learning Machines. 3 ed. Englewood Cliffs, NJ: Prentice Hall; 2009, 916 p.
[11] Cybenko G. Aproximation by superpositions of a sigmoidal function. Math Control Signals Syst 1989;2:303–14.
[12] Funahashi K. On the approximate realization of continuous mappings by neural networks. Neural Net 1989;2:183–92.
[13] Battiti R. First and second order methods for learning: between steepest descent and Newton's method.. Neural Comput 1992;4(2):141–66.
[14] Hagan MT, Menhaj M. Training multilayer networks with the Marquardt algorithm. IEEE Trans Neural Net 1994;5(6):989–93.
[15] Singhal S, Wu I. Training multilayer perceptrons with the extended Kalman filter. Adv Neural Inf Process Syst 1989;1:133–40.
[16] Iiguni Y, Sakai H, Tokumaru H. A real time learning algorithm for a multilayered neural network based on the extended. IEEE Trans Signal Process 1992;40 (4):959–66.
[17] Skorohod B. Diffuse Initialization of Kalman Filter. J Autom Inf Sci 2011;43 (4):20–34. Begell House Publishing Inc, USA.
[18] Skorohod B. Diffusion learning algorithms for feedforward neural networks. Cybernet Syst Anal May, 2013;49(3). Publisher: Springer New York.
[19] Kim C-T, Lee J-J. Training two-layered feedforward networks with variable projection method. IEEE Trans Neural net February, 2008;19(2).
[20] C.-T. Kim, J.-J. Lee and H. Kim, Variable projection method and Levenberg-Marquardt algorithm for neural network training. In: IEEE Industrial Electronics, IECON 2006-32nd Annual Conference on 6–10 Nov. 2006. p. 4492–4497.

[21] Skorohod B. Learning algorithms for neural networks and neuro-fuzzy systems with separable structures. Cybernet Syst Anal 2015;51(2):173−86. Publisher: Springer New York.
[22] Huang GB, Wang DH, Lan Y. Extreme learning machines: a survey. Int J Mach Lean Cybernet 2011;2(2):107−22.
[23] Broomhead D, Lowe D. Multivariable functional interpolation and adaptive networks. Complex Syst 1988;2:321−55.
[24] Simon D. Training radial basis neural networks with the extended Kalman Filter. Neuro Comput 2002;48:455−75. October 2002.
[25] Takagi T, Sugeno M. Fuzzy identification of systems and its application to modeling and control. IEEE Trans Syst Man Cybern. 1985;SMC-15:116−32.
[26] Simon D. Training fuzzy systems with the extended Kalman filter. Fuzzy Sets Syst 2002;132(2):189−99.
[27] Goddard J, Parrazales R, Lopez I, de Luca A. Rule learning in fuzzy systems using evolutionary programs. IEEE Midwest Symp Circuits Syst 1996;703−9. Ames, Iowa.
[28] Magdalena L, Monasterio-Huelin F. Fuzzy logic controller with learning through the evolution of its knowledge base, Internat. J Approx Reason 1997;16:335−58.
[29] Nelles O. Nonlinear System Identification. From Classical Approaches to Neural Networks and Fuzzy Models. Berlin: Springer; 2001. p. 785.
[30] Bruls J, Chou CT, Haverkamp BRJ, Verhaegen M. Linear and non-linear system identification using separable least-squares. Eur J Control 1999;5:116−28.
[31] Edrissi H, Verhaegen M, Haverkamp B, Chou C.T. Off- and On-Line Identification of discrete time using Separable Least Squares. In: Proceedings of the 37th IEEE Conference on Decision & Control, Tampa, FL.1998.
[32] Stubberud SC, Kramer KA, Geremia JA. Online sensor modeling using a neural Kalman filter. IEEE Trans Instrum Meas 2007;56(4):1451−8.
[33] Stubberud SC, Kramer KA, and Geremia JA. System identification using the neural-extended Kalman filter for state-estimation and controller modification. In: Proceedings of the International Joint Conference on Neural Networks, 2008. p. 1352−1357.
[34] Kral L, Simandl M. Neural networks in local state estimation. Methods and Models in Automation and Robotics (MMAR), In: 17th International Conference on 27−30 Aug 2012. p. 250−255.
[35] Jin L, Nikiforuk P, Gupta M. Approximation of discrete-time state-space trajectories using dynamic recurrent neural networks. IEEE Trans Autom Control 1995;40(40), N.7.
[36] Hagan MT, Demuth HB, De Jesús O. An introduction to the use of neural networks in control systems. Int J Robust Nonlinear Control 2002;12(11):959−85, September.
[37] Ljung L, Soderstrom T. Theory and Practice of Recursive Identification. Cambridge, MA: MIT Press; 1983.
[38] Haykin S. Adaptive Filter Theory. Englewood Cliffs, NJ: Prentice-Hall; 1991.
[39] Mosca E. Optimal, Predictive and Adaptive Control. Englewood Cliffs, NJ: Prentice Hall; 1995.
[40] Bellanger MG. Adaptive Digital Filters and Signal Analysis. New York: Marcel Dekker; 1987.
[41] Cioffi JM, Kailath T. Fast, recursive-least-squares transversal filters for adaptive filtering. IEEE Trans Acoust Speech Signal Process. 1984;32(2):304−37.
[42] Hubing NE, Alexander ST. Statistical analysis of initialization methods for RLS adaptive filters. IEEE Trans Signal Process. 1991;39(8):1793−804.
[43] Eom K-S, Park D-J. Analysis of overshoot phenomena in initialisation stage of. RLS algorithm, ELSEVIER. Signal Process, 44. 1995. p. 329−39.
[44] Moustakides GV. Study of the transient phase of the forgetting factor RLS. IEEE Trans Signal Process 1997;45:2468−76.

[45] Skorohod B. Asymptotic of linear recurrent regression under diffuse initialization. J Autom Inf Sci 2009;41(5):41–50. Begell House Publishing Inc, USA.

[46] Skorohod B. Oscillations of RLSM with diffuse initialization. Automation of processes and control: proceedings of SSTU, 146, P. 40–45, Sevastopol, 2014 (in Russian).

[47] Albert A. Regression and the Moore-Penrose Pseudoinverse. New York: Academic Press; 1972. 177 p.

[48] Albert A, Sittler R. A method for computing least squares estimators that keep up with the data. SIAM J Control 1965;3(3): 384, 417.

[49] Stoica P, Ashgren P. Exact initialization of the recursive least-squares algorithm. Int J Adapt Control Signal Process 2002;16(3):219–30.

[50] Ansley CF, Kohn R. Estimation, filtering and smoothing in state space models with incompletely specified initial conditions. Ann. Statist. 1985;13:1286–316.

[51] Koopman SJ. Exact initial Kalman filtering and smoothing for non-stationary time series Models. J Am Stat Assoc 1997;92(440):1630–8.

[52] De Jong P. The diffuse Kalman Filter. Ann. Statist. 1991;19:1073–83.

[53] Kwon WH, Kim PS, Park P. A receding horizon Kalman FIR filter for discrete time invariant systems.. IEEE Trans Autom Control 1999;44(9):1787–91.

[54] Kwon WH, Kim PS, Han SH. A receding horizon unbiased FIR filter for discrete-time state space models. Automatica. 2002;38(3):545–51.

[55] Harvey AC, Pierse RG. Estimating missing observations in economic time series. J Am Stat Assoc., 79. 1984. p. 125–31.

[56] Bar-Shalom Y, Rong X, Kirubarajan T. Estimation with Applications to Tracking and Navigation. New York: Johh Wiley and Sons; 2001.

[57] Gantmaher FR. Matrix Theory, M.: Fizmathlit, 2004, 560 p.

[58] Narendra KS, Parthasarathy K. Identification and control of dynamical systems using neural networks. IEEE Trans Neural Net. Mar. 1990;1(1):4–27.

[59] Jinkun L. Radial Basis Function (RBF) Neural Network Control for Mechanical Systems. Heidelberg: Springer; 2013.

[60] Jazwinski AH. Limited memory optimal filtering. In: Proceedings 1968 Joint Automatic Control Conference. Ann Arbor, MI, 1968, p. 383–393.

[61] Hyun KW, Ho LK, Hee HS, Hoon LC. Receding horizon FIR parameter estimation for stochastic systems. Int Conf Control Autom Syst 2001;1193–6.

[62] Alessandri A, Baglietto B, Battistelli G. Receding-horizon estimation for discrete-time linear systems. IEEE Trans Autom Control 2003;48(3):473–8.

[63] Wang Z-o, Zhang J. A Kalman filter algorithm using a moving window with applications. Int J Syst Sci 1995;26(9):1465–78.

[64] Gustafsson F. Adaptive filtering and change detection. Chichester: Wiley; 2001.

[65] Hoerl A, Kennard R. Ridge regression: biased estimation for northogonal problems. Technometrics 1970;12:55–67.

[66] Korn A, Korn T. Mathematical Handbook for Scientists and Engineers. New York: McGraw-Hill Book Co; 1961. p. 943.

[67] Seber G. A Matrix Handbook for Statisticians. Hoboken, NJ: John Wiley & Sons, Inc; 2007, 559 pp.

[68] Cook D, Forzani L. On the mean and variance of the generalized inverse of a singular Wishart matrix. Electron J Stat 2011;5:146–58.

[69] Wimmer H.R. Stabilizing and unmixed solutions of the discrete time algebraic Riccati equation. In: Proceeding Workshop on the Riccati equation in Control, Systems, and Sygnals. 1989. p. 95–98.

[70] Polyak BT. Introduction to optimization. Translations Series in Mathematics and Engineering. New York: Optimization Software Inc. Publications Division; 1987.

[71] Ravi P. Agarwal Difference Equations and Inequalities: Theory, Methods, and Applications. Boca Raton, FL: CRC Press; 2000. p. 971.
[72] Johnstone RM, Johnson CR, Bitmead RR, Anderson BDO. Exponential convergence of recursive least squares with exponential forgetting factor, Decision and Control, 1982 21st IEEE Conference. p. 994–997.
[73] Harville D. Matrix Algebra From a Statistician's Perspective. Berlin: Springer Science & Business Media; 2008, 634 pp.
[74] Lakshmikantham V, Trigiante D. Theory of Difference Equations—Numerical Methods and Applications. 2nd ed. New York: Marcel Dekker; 2002, 300 p.
[75] Wasan MT. Stochastic Approximations. Cambridge: Cambridge University Press; 1969.
[76] Bertsekas Dimitri P. Incremental least squares methods and the extended Kalman Filter. SIAM J Optim. 1996;6:807–22.
[77] Curve Fitting Toolbox 3. MathWorks, Inc.
[78] Fuzzy Logic Toolbox. The MathWorks, Inc.
[79] System Identification Toolbox. The MathWorks, Inc.
[80] Bezdek J, Keller J, Krishnapuram R, Kuncheva, Pal L. Will the real Iris data please stand up? IEEE Trans Fuzzy Syst 1999;(7):368–9.
[81] Anderson BDO, Moore J. Optimal Filtering. Englewood Cliffs, NJ: Prentice Hall; 1979.
[82] Jazwinski AH. Stochastic Processes and Filtering Theory. New York: Academic Press; 1970.
[83] S.H. Han, P.S. Kim and W.H. Kwon Receding horizon FIR filter with estimated horizon initial state and its application to aircraft engine systems. In: Proceeding of the 1999 IEEE international conference on Control Application Hawai, August 22–27, 1999.
[84] Narendra KS, Parthasarathy K. Gradients methods for the optimization of dynamic systems containing neural networks. IEEE Trans Neural Net 1991;2(2):252–62.
[85] Sum J. Extended Kalman Filter Based Pruning Algorithms and Several Aspects of Neural Network Learning. PhD Dissertation. Hong Kong: Department of Computer Science and Engineering, Chinese University of Hong Kong; 1998.
[86] http://www.festo-didactic.com/int-en/learning-systems/education-and-research-robots-robotino/robotino-workbook.htm.
[87] Mayergoyz ID. Mathematical Models of Hysteresis and Their Applications. Amsterdam: Elsevier Series in Electromagnetism; 2003.
[88] Advances in neural networks in computational mechanics and Engineering by Ghaboussi J. in Advances of soft computing in engineering, Springer, p191, 2010, 470 p.
[89] Yun GJ, Yun, Ghaboussi J, Amr S. Modeling of hysteretic behavior of beam-column connections based on self-learning simulation, Report, Illinois, August 2007, 224 p.
[90] Ghaboussi J, Garrett JH. and Wu X. Material modeling with neural networks. In: Proceedings of the International Conference on Numerical Methods in Engineering: Theory and Applications. Swansea; 1990.
[91] Ghaboussi J, Garrett JH, Wu X. Knowledge-based modeling of material behavior with neural networks. J Eng Mech Div 1991;117:132–53.
[92] Dash PK, Panda SK, Liew AC, Mishra B, Jena RK. A new approach to monitoring electric power quality. Elect Power Syst Res 1998;46:11–20.
[93] Rechka S, Ngandui E, Jianhong X, Sicard P. Analysis of harmonic detection algorithms and their application to active power filters for harmonics compensation and resonance damping. Can J Elect Comput Eng 2003;28:41–51.
[94] Dash PK, Swain DP, Liew AC, Rahman S. An adaptive linear combiner for on-line tracking of power system harmonics. In: IEEE Trans Power Appar. Syst. 96WM 181–8 PWRS, January 21–25, 1996.

INDEX

Note: Page numbers followed by "*f*" and "*t*" refer to figures and tables, respectively.

M

N

O

P

R

S

W

Printed in the United States
By Bookmasters